SCENES OF ATTENTION

SCENES OF ATTENTION

ESSAYS ON MIND, TIME, AND THE SENSES

EDITED BY

D. GRAHAM BURNETT
AND
JUSTIN E. H. SMITH

Columbia University Press   *New York*

Columbia University Press
Publishers Since 1893
New York   Chichester, West Sussex
cup.columbia.edu

Copyright © 2023 Columbia University Press
All rights reserved

Library of Congress Cataloging-in-Publication Data
Names: Burnett, D. Graham, editor. | Smith, Justin E. H., editor.
Title: Scenes of attention : essays on mind, time, and the senses /
D. Graham Burnett and Justin E. H. Smith, eds.
Description: New York : Columbia University Press, 2022. |
Includes bibliographical references and index.
Identifiers: LCCN 2023014412 | ISBN 9780231211185 (hardback) |
ISBN 9780231211192 (trade paperback) | ISBN 9780231558785 (ebook)
Subjects: LCSH: Visual perception. | Attention.
Classification: LCC BF241 .S336 2023 |
DDC 152.14—dc23/eng/20230418
LC record available at https://lccn.loc.gov/2023014412

Cover design: Chang Jae Lee
Cover image: © Shutterstock

CONTENTS

Introduction: Thinking Attention 1
D. GRAHAM BURNETT AND JUSTIN E. H. SMITH

I HISTORIES OF ATTENTION

1 The Discovery of Attention 23
RICHARD J. SPIEGEL

2 Attention and Boredom in Early American Psychology 44
HENRY M. COWLES

3 Paying Attention to the Birds: Ornithologists and Listening 63
ALEXANDRA HUI

4 Attention, Art, and Psychotherapeutics 81
JULIAN CHEHIRIAN

II PHILOSOPHIES OF ATTENTION

5 Attention: Mechanism and Virtue 103
CARLOS MONTEMAYOR

6 Attention, Technology, and Creativity 124
CAROLYN DICEY JENNINGS AND SHADAB TABATABAEIAN

7 Attending to Absence, and the Role of the Imagination 142
JONARDON GANERI

8 Dispatch from the Jhāna Wars: Attention Practice in Online Buddhism 160
JOHN TRESCH

III ATTENTION, TECHNOLOGY, AND CULTURE

9 Wearable Attention: Course Corrections for Wandering Minds 187
NATASHA DOW SCHÜLL

10 Attentional "Ownership": Online Education and Self-Possession 212
BRIAN YUAN

11 Attention Is All You Need: Humans and Computers in the Time of Neural Networks 230
NICK SEAVER

12 Medium Focus 249
JOANNA FIDUCCIA

IV ENDGAME(S)

13 Attention Fast, Attention Slow: Obsession, Compulsion, and Holding Close 275
YAEL GELLER

14 Units of Intensive Care: Poetic Attention and the Precarious Body 291
LUCY ALFORD

Bibliography 317
List of Contributors 343
Index 349

SCENES OF ATTENTION

INTRODUCTION

Thinking Attention

D. GRAHAM BURNETT AND JUSTIN E. H. SMITH

A SCENE OF ATTENTION: THINKING IN CIRCLES

Mr. Sanders was a fundamentally unusual teacher. Fourth grade for me, Graham. In Indiana, where I grew up, somewhere outside Indianapolis. A small Catholic school. It would have been 1979 or so. I think the place was called St. Luke's.

Mr. Sanders had a boa constrictor. And that was cool. He brought it into school sometimes and taught math with the thing around his neck.

He also had a genuine switchblade. An old-fashioned pigsticker (as my father called it) like the ones that James Dean's character and his nemesis use in the observatory fight in *Rebel Without a Cause*. It is some measure of the distance between 1979 and our moment that Mr. Sanders kept the knife in the top drawer of his desk and used it for ordinary fourth-grade-teacher activities—like getting masking tape off the door or tightening a screw on the overhead projector.

In that same top drawer Mr. Sanders additionally possessed a heavy metal ruler (it may have been brass or some sort of tinted aluminum) that he called the "Golden Rule," the purpose of which was to whack the ass of a child spread-eagled with his/her hands on the chalkboard. I saw two boys in that position in the course of fourth-grade. Each was given the option of writing "I will not [*whatever annoying thing he had done*]" two hundred times on the chalkboard after school in lieu of three whacks with the Golden Rule—they both took the text-based option.

Finally (and it is my strongest memory of fourth grade), Mr. Sanders could rip a telephone book in two. I still am not 100 percent sure how he did this. I suppose I could, today, look on the Internet and discover the trick. But I wish to preserve the sense of mystery that inheres in my naïve recollection of the act. It was a beefy Yellow Pages, and it took him quite a while to do it. There was a good deal of theater in the performance. He would leap into the air and give a wild yelp, snap his wrists as he torqued on the volume, and then demonstrate his progress by showing the class the fissure that gradually opened in the thick sheaf. Let me be clear: I am *not* saying he tore the book in two by severing it at the binding, such as one might tear a giant paperback novel into a "part one" and a "part two" in order to pack light on a short trip. No. I am saying he tore *through the actual pages*. All of them. The result was more like two halves of a sandwich. If it had been a novel, every line on every page would have been split by a manual caesura. We were, understandably, rapt by this display of supernatural strength.

I liked Mr. Sanders a great deal. We all did. He was a bit fierce, but he was a self-possessed and dynamic character, and he had a definite knack for holding the attention of a room of fourth graders. His general burly truculence translated into an energy that said, "I've rolled up my sleeves, and you're gonna learn long division." He came on like the lead edge of a hurricane, sweeping everything up into a tornado of arithmetic.

It is interesting for me now, in my adult person, to attempt a reconstruction of the life-world of Mr. Sanders. Knowing him only as a figure from my youth, there is something slightly vertiginous about the work of considering his full humanity, such as I can now imagine it from my current perspective. For instance, I can now see that . . . well, it was the 1970s. And Mr. Sanders was a product of that age in a number of respects. The wide tie, for instance. The longish black hair combed down around his balding pate. The experiments in "remote sensing" that he ran on us as a class.

Wait. What?

Yes. That last bit may sound as unsettling as the switchblade or the tender farm boys "assuming the position" at the front of the class. But I am again at pains to emphasize that I, anyway, experienced it all without a hint of trauma. (As far as I know, anyway.)

On the contrary, I liked all this a great deal better than the public school I had attended the year before, where I had been genuinely miserable. (My mother knew something was wrong when I began compulsively designing and sketching various fantastic "torture chambers" in my school notebooks, most based on a conveyor-belt system that ran a daisy chain of children sequentially through shark tanks, meat grinders, boiling oil, etc.).

So, the remote sensing experiment. It went like this: one day Mr. Sanders said that we were going to do something different for science class. The parameters of the enterprise were sketched out. One of us was going to be sent to the gymnasium to sit there for twenty minutes with eyes closed. This person was the "receiver." The rest of us, including Mr. Sanders, were going to stay in the classroom. We were going to configure our desks so that we were all facing the gymnasium. Then we were all going to close our eyes and put our heads down and collectively labor, in silence, to *convey to the receiver a telepathic thought.* This thought/object/word thing would be determined by Mr. Sanders after the departure of the receiver, and we would all endeavor, together, to send this particular "percept" (as the modern psychologist would call it) to the receiver in the gymnasium—which was, as I recall, about fifty meters down the hall.

Again, in retrospect this seems quite mad. If my kid came home saying that this was what she was doing in science class, I would blow a gasket. But it was a different time and a different place. Indiana. In the 1970s. And frankly, my recollection is that this activity was undertaken in a promising spirit of genuine inquiry. We were up for it.

A volunteer was solicited, and this student headed down the hall to set up in the gymnasium with his internal receptors on the qui vive. The rest of us applied ourselves to the noisy task of orienting our integrated desk-chair contraptions toward the distant gym. Mr. Sanders announced that the datum we would undertake to convey would be a *square.* I don't recall his offering a lot of instructions as to how we were to go about sending the mental image of a square (or just the word?) through the diaphanous ether of the real, but his let's-*do*-this-thing energy was palpable. He swung his own chair brusquely into position with a loud thump and barked, "Eyes closed and heads down! Squares!"

I closed my eyes.

It is that internal condition I wish to try to recover now. Since it remains, for me, a kind of primal scene of a certain kind of mental work—a primal *scene of attention* of the sort with which this book is concerned. I am not sure that, until that point in my life, anyone had ever asked me to close my eyes and do sustained, durational, mental "work" of any kind—other than prayer, of course. But prayer is a special category. This was school. This was *science* class. This was a formal "exercise" that I understood was supposed to unfold without reference to supernatural agencies of any obvious sort. We were trying to figure out whether thought could travel through space, in something of the way that sound does.

So I sat there. With my eyes closed.

But the fact of the matter is that at the age of nine, I did not in any way believe that thought was capable of a direct conveyance like this. I have no idea why I was so absolutely certain of the absurdity of the idea. After all, I believed in God and angels and souls, and intercessory prayer and grace, too. So it stands to reason that I might have been open to the notion of the immaterial propagation of propositional content from one human being to another across empty space. But the idea seemed risible to me.

I was a *very* obedient child. And very-very determined not to displease my elders in any way. Which I mention because this fact about my character will assist the reader in understanding just how forcefully I dissented from the fourth-grade telepathy experiment. Despite my general "goody-good" desire to please teachers of all kinds, and in the face of my specific enthusiasm for Mr. Sanders, I decided, sitting there with my head down on my desk, staring into the darkness of my mental space, that I would think, with all my power, *circles*.

A moment on this act of defiance is perhaps in order.

After all, a perfect dissent from the experiment would possibly have involved a kind of contemptuous indifference to the whole business. Why, if I disbelieved in the possibility of telepathic communication, did I choose to engage in a kind of telepathic *sabotage*, instead of just tuning out altogether?

Frankly, I still find this puzzling. My introspective efforts to reconstruct the logic of my motive have led me to a sense that I wanted to prove that it did not matter what one thought, because thoughts did not travel in this way. Although a better experimental technique for falsifying remote sensing would presumably have been to think squares along with everyone else and then have my classmate come back from the gymnasium and announce that iguanas or some such random thingamabob had popped into his head. I don't know why I didn't do that.

What I did do, anyway, was think circles. And I thought of them *conveyingly*. Which is to say that I set my mind forcefully to the task of visualizing circles and attempting to "send" those circles into the world in a radiating fashion, such that they might be received by others. But let me emphasize (paradoxically? perversely?) that I did this as a protest, because I radically disbelieved in the efficacy of such a practice. Or at least I thought I did. Though it is worth asking why I "did" it as a way of (not?) showing my disbelief.

And then—I swear by all that is precious that this really happened!—Mr. Sanders piped up, his stern voice entering the orbicular miasma of the darkened chamber of my mind: "Somebody in this room is thinking *circles*!"

HISTORICIZING THE ATTENTIVE PERSON

This book takes up the problem of attention, and it does so from a diverse range of disciplinary perspectives. Authors of the chapters that follow come from anthropology, philosophy, comparative literature, and the history of science; they are medical doctors, poets, artists, and art historians. They broadly share a concern with what the French theorist Yves Citton calls the "ecology" of attention: the richly matrixed, time-bound, reciprocating, recursive, and ultimately *dynamic* nature of our attentive lives. All the contributors are in various ways preoccupied by technology. And all are in various ways mindful of the rise of our "attention economy"—and of the unique challenge of thinking attention in a world increasingly focused on the intensive "fracking" of human persons for the vaporous monetary value of their eyeballs and earholes. But each of our authors thinks of "attention" *differently* in these pages: we have cats in cages, a wooden box containing a hidden speaker, neo-Kantian schoolmasters, a missing friend, the artwork of people with mental illness, bloodless neural networks, two ornithologists, and a mother in a coma. As editors, we sometimes felt that our role was that of Mr. Sanders: we encouraged everyone to orient their desks in the same direction, close their eyes, and think *attention*. And it was you, reader, that we all had in mind. We hoped to send you our thinking, remotely—by means of these sewn signatures.

Scenes of Attention is our title. It is our method, too. Like this introduction, each chapter begins with a "scene" of attention. Our wager here has its origins in the brilliant introductions to two wonderful books: Jeff Dolven's *Scenes of Instruction in Renaissance Romance* and Martin Puchner's *The Drama of Ideas*. Each of these studies centers a theoretical inquiry on a certain dogged commitment to scenographic specificity. Following those leads, we asked our authors to work from (and with) discrete "dramaturgies" of attention: concrete situations, bound in time and space, inhabited by people who are trying (but often "failing") to *attend*. Some authors took up this challenge with gusto, working a given attentional situation with close phenomenological scrutiny. Others touched their scenes lightly, leaving them as a kind of bookplate on the endpaper of their inquiry. But each has been faithful to the conceit, and their scenes, taken together, give scope and ambit to the volume—even as they return us, again and again, to the concrete world of our chosen problem.

And our chosen problem is indeed concrete—not to mention urgent. Why does the subject of attention seem to cut so close to the stuff of our very persons? Why does it engender such intense debates? Why have we been in one or another condition of "attentional crisis" for more than a century?

One way of answering questions like these has been to reach for a litany of dramatic changes in the technological conditions of modernity. The primordial crisis of modern attention lies in the second half of the nineteenth century, the period that saw the rise of the steam engine, intensified factory labor, the rapid expansion of daily newspapers (and hence increasingly fast-paced news cycles), and, of course, railway travel—motion through space at speeds previously reached only by people who threw themselves out of buildings.

The new ubiquity of speed precipitated a new anxiety about the way everything was continuously slipping by. Through a train window (see Yael Geller's contribution in chapter 13), the world disappeared as fast as it appeared, and the idle passenger experienced a new kind of smooth and speedy transit of all things on the other side of that transparent screen. One can discern, in sensitive writings from the period, that many commentators felt a novel and strange discomfort with the way people, views, images, and places were whisked away by what truly seemed to be a general increase in the pace of existence. This sense of acceleration is inseparable from what we think of as "the modern," and these dynamics (an experience of what feels like an ever-increasing *speed of life*) characterize a wide range of human expression in many parts of the world going back a surprisingly long time. Something real has happened in this respect. And it is not impossible that a major driver of our anxiety about attention (meaning anxiety about our ability to give our mind and senses to what need their time and work) has simply been a function of this pervasive experience of acceleration. As things move faster, more stuff comes before us in the same amount of time. The language of "information overload" starts to feel relevant, and the language of "attention" becomes a way of expressing concern about filtering the flows. In this context, "attention" functions as a tool for *managing the excess*; to "be attentive" is to be able to hold on to what needs to be held, and to achieve this under conditions in which it feels increasingly difficult to do exactly this.

Many historians feel uneasy about this kind of explanation. It smacks of something called "technological determinism." This is generally seen as a bad thing among academic students of history—though, as a matter of practice, it is widely accepted by just about everyone else as a basic feature of the way the world works. The technological determinist believes that changes in technology "determine" historical change—meaning they "cause" those changes, the way that one billiard ball hitting another "causes" that second ball (under ordinary conditions) to roll; everything you need to know about the rolling of the second ball is "in" the first ball, and what is going to happen can be calculated in advance (pretty much).

The paradigmatic technological determinist would seem to think that society is a bit like that second ball—and that a given "technology" works like the first one. So, somebody invents a fast-moving locomotive and—*bonk!*—people start having weird ideas about attention. Maybe that example is a little on the abstract side. But you get the idea: technology X causes social effect Y. We are surrounded by this way of thinking (e.g., our phones are making us more isolated, the internet is making us more stupid—or maybe smarter). And it does seem like there is something to these sorts of arguments. The objection, however, is that *human beings make the technologies*—and they make them in the context of other human beings needing and wanting various things (see Natasha Dow Schüll's contribution in chapter 9 and Nick Seaver's in chapter 11). In this sense, the technologies are the *effects* (rather than the causes); hence, one might do better, in examining changes in society, to focus not on the "causal" role of the technologies (as if metal contraptions had some sort of independent agency or came from outer space) but rather on the larger matrix of human need and aspiration and fantasy and fear that gave rise to the technologies in the first place.

Those of you who smell a chicken-and-egg type of situation here—well, there probably is some of that. Still, given how ubiquitous and "easy" technological determinist arguments seem to be, I would say there are grounds for checking yourself (and others) every time they come up. If you dig down under the dynamics of a technology that appears to be changing society, can you spot the sociocultural drivers that led to the rise of the new technology in the first place? This is a good exercise in almost any situation. The reality is that new ideas and new beliefs can be hugely powerful agents of historical change. But they are harder to point to than trains and phones, and ideas are hard to put in a museum.

There is, in fact, a different account of the attentional crises of the modern period: one that is much more concerned with conceptual, rather than technological, change. It goes like this: the nineteenth century saw a breakdown (or was it just a transformation?) of some of the deepest ways of thinking about a human person. Painting with a broad brush, the period saw a profound destabilization of traditional ideas about the spiritual and physical aspects of being (see Richard Spiegel's contribution in chapter 1, as well as Jonathan Crary's important work, *Suspensions of*

INTRODUCTION ❧ 11

Perception). For many people, especially those associated with university learning and urbane sophistication in Western Europe (and its colonial/imperial metastases), religious notions of the unique and unitary "soul" of every individual lost much of their traction. Relatedly, shifts in the understanding of the complexity of human perception and cognition led to widespread defection from "classical" models of subjectivity.

What might that mean? What is the classical model of the subject?

One way to think of it is to use a favorite Enlightenment metaphor: the camera obscura—that pinhole-camera-like device through which the outside world is projected as an image on the wall of a dark little box.[1] The camera obscura impressively captured something real about the working of the human eye (and animal eyes, too). And a general visualist bias made the eye feel like a pretty good model for the working of the mind as a whole. There was a world out there. And, conveniently, something pretty much like that same world shaped up inside us as a kind of "mapping" produced by our senses. Sure, there could be mess-ups in that mapping (bad eyesight, hysterical illusions), but if you were sane, rational, and not intoxicated (and, ideally, a propertied White male with a walking stick, standing up nice and tall)—that is, if you were a "classical subject"—your head was like an ideal camera obscura: your senses were continuously generating a nice, accurate, detailed "map" of everything around you all the time. Maintained inside your head, this conformal representation permitted you to orient, assess, and act in the world. A happy confidence in geometry and a general sense of the mathematical armature of all things helped give this image of the eye-mind a certain robust plausibility.

But where were "you" in this mind-movie-theater? Well, "you" were, in some nebulous sort of way, at the center of it all. The whole show was "for" you, and "you" were there as the central and culminating spirit-spectator of the spectacle. To be sure, different thinkers reasoned in different ways about the exact locus and character of the "self" in relation to this general model of sensory existence, but the model itself was broadly shared.

Over the nineteenth century, however, that model began to break down. The stable vision of vision itself as essentially *stable*—and of the human subject as a kind of anchored, Archimedean eye surveying what obtains with a certain patrician sovereignty/autonomy—ceased to satisfy across a midcentury watershed. Why? One can offer several accounts. Certainly the empirical study of vision and the other sensory modes revealed strange and oozy facts. For instance, people did not, as it turns out, just "see" in the way that a camera obscura "sees." Yes, the physical eye might be a little pinhole camera, where an optical image formed on the retina, but "seeing" happens somewhere in a large wet organ inside the head, and various psychophysical experiments in this period demonstrated that what people "saw" (and heard and felt) was in fact a function of all kinds of things that had nothing to do with the geometry of tracing rays of light through the lens of the eye. Even the most "rational" and precise observers (astronomers, say) could be shown to differ in their observations in weird and unsettling ways that seemed to have to do with nebulous stuff like "personality" (some people are jumpy and expectant, some are dogged and skeptical; such attributes directly affect how perception works). New preoccupations with optical illusion and visual deception marked a growing belief that vision did not reinforce the unity and perspectival coherence of the subject, but rather indexed its fragmentation.

If this story about the collapse of the classical subject is right (and there are reasons to think it is at least *not wrong*), it sets the stage for a very interesting interpretation of the rise of new anxieties around attention in the same period. Since, by these lights, it can suddenly appear as if the whole discourse of attention might best be understood as a new language within which to attempt, in the wake of the failure of the classical model, a rearticulation of some kind of unitary or coherent subject. Within what was increasingly perceived as a fractured, distributed, and inscrutably somatic sensory field, concern with attention was concern with the actual locus of the stable, accountable *subject* within the blooming, buzzing confusion of the manifold (see Julian Chehirian's contribution in chapter 4 for a close look at how psychoanalysis emerged out of and worked with this problem—*in attentional terms*). In this sense, the sharp irruption of explicit fretting about attention in the second half of the nineteenth century may well tell us not that people who had been good at "paying attention" before trains and factories and electric lights were now having trouble doing so after their introduction, but rather that thinking of the problems of human persons (in labor, education, and politics) as *attention problems* reflected an effort to feel around for exactly where a "human person" might be.[2] Attention, understood this way, represented a new and powerful conceptualization of the human subject.

It is a mode of articulation of the human subject that remains with us (see Jonardon Ganeri's contribution in chapter 7) and remains powerful—newly powerful, in fact, because a kind of *industrial revolution* in the advertising industries, in conjunction with a radical intensification in digitally mediated lifeways, has brought us a "surveillance capitalism" rooted in hypercommodified attention. Authors including Tim Wu, James Williams, and Shoshana Zuboff have heightened our collective awareness of this brave new world, and activist groups like Time Well Spent and the Friends of Attention (with which both of us have been involved) have worked to theorize and promote forms of attention resistant to financialization.³ But the way forward is in no way obvious. This book was conceived in the context of mounting concern about exactly these dynamics, and while our volume can hardly be said to be "on the barricades" in the fight for what the Friends of Attention call the "Attention Liberation Movements" (or *ALMS*), this is a book that is meant to deepen the thinking needed to protect the freedom of attention, and the forms of life out of which it emerges (and which it makes possible).⁴

A BRIEF SKETCH OF THE PAGES AHEAD

Our volume is organized into four parts, the first three of which each contain four chapters. The final part, "Endgame(s)," contains a little suite of valedictions. Contributions to part I, "Histories of Attention," focus principally on key moments in the history of the study, direct or indirect, of attention. In chapter 1, "The Discovery of Attention," the historian Richard J. Spiegel traces the development of reflection on attention in German philosophy and incipient anthropology, from Immanuel Kant through Ernst Platner and Johann Herbart, revealing the surprising centrality of attention in theoretical reflections about the mind in the German Enlightenment. In chapter 2, "Attention and Boredom in Early American Psychology," the historian Henry M. Cowles continues the historical development through to the end of the nineteenth century, considering in particular the unique role that research on the affective state known as boredom—a condition both distant from attention but also surprisingly revelatory of some of its features—played in the development of experimental psychology in the United States. Curiously, one of the key figures in this story is a psychologist by the name of Edwin Boring. In chapter 3, "Paying Attention to the Birds: Ornithologists and Listening," the historian of science Alexandra Hui shifts our attention to an undertheorized sensory modality: the auditory. She shows how, in a field such as ornithology, in which one often has no choice but to let one's ears take the lead, it sometimes happens in collaborative research that forms of attentive listening are conveyed to others, creating a sort of "phenomenological loop" and standardizing patterns of attention within the collaboration. In chapter 4, "Attention, Art, and Psychotherapeutics," the historian of the mind sciences Julian Chehirian investigates another important strain of psychological theorizing in the late nineteenth and early twentieth centuries, namely, psychoanalysis. Examining in particular the emergence of a practice of "free-floating attention" out of Sigmund Freud's appropriation from literature of the idea of "free association," Chehirian traces a gradual shift in some strains of psychoanalytic theory and practice toward a relatively greater preoccupation with nonverbal expressivity and shows how this preoccupation shaped the "art therapy" movement of the postwar period.

Part II, "Philosophies of Attention," focuses on attention as a philosophical problem. It begins with a chapter from the philosopher Carlos Montemayor entitled "Attention: Mechanism and Virtue," exploring the fundamental relationship between attention as a cognitive state and attention as a moral virtue. Adopting an approach at the intersection between philosophy and psychology, Montemayor argues that attention in its ideal expression entails autonomy, and thus falls within the domain of moral philosophy, while also arguing that this ideal is particularly threatened in the technological circumstances of the present moment. In chapter 6, "Attention, Technology, and Creativity," the philosophers Carolyn Dicey Jennings and Shadab

Tabatabaeian draw on neuroscience, experimental psychology, and conceptual analysis to examine the relationship among attention, technology, and creativity. Jennings and Tabatabaeian seek to measure the impact that recent digital technologies have had not just on attention in general (a commonly thematized problem), but specifically on attention as a precondition of creative work. Marshaling the familiar distinction between "top-down" and "bottom-up" attention (more classically expressed in terms of the "voluntary" and "involuntary," respectively), the authors investigate the various respects in which recently developed digital technologies undermine our capacity for top-down attention, which is crucial for creativity. In chapter 7, "Attending to Absence, and the Role of the Imagination," the philosopher Jonardon Ganeri elucidates the theory of negative attention developed by the Bengali philosopher Krishnachandra Bhattacharyya in the early twentieth century and shows the significant links it has with some near-simultaneous developments in existential phenomenology, notably in Jean-Paul Sartre's reflections on the phenomenology of absence. In the final chapter of part II, "Dispatch from the Jhāna Wars: Attention Practice in Online Buddhism," John Tresch investigates a fascinating chapter at the intersection of philosophy, religion, and global cultural exchange, namely, the reception and transformations of Buddhist meditational practices in the West, and especially the increasing pace of these transformations following their uptake into new digital platforms. Writing from the perspective of a participant observer, Tresch is particularly interested in charting the emergence of a peculiar new culture that fuses the ancient practice of jhāna with contemporary technoscience and places a high premium on the cultivation of what it calls "mindfulness," a value that is in many respects specific to our contemporary technological and economic reality.

In part III, "Attention, Technology, and Culture," our authors approach the problem of attention from perspectives drawing on anthropology, media studies, and art history. In chapter 9, "Wearable Attention: Course Corrections for Wandering Minds," the anthropologist Natasha Dow Schüll investigates recent prosthetic technologies intended to enhance attention through a brain–machine interface. Exploring the limits of such technological enhancements while avoiding the temptation to praise or condemn, she seeks to determine, most importantly, what new sorts of attentional subjects these devices may bring into being. In chapter 10, "Attentional 'Ownership': Online Education and Self-Possession," the anthropologist Brian Yuan offers a close study of the Summit Learning Platform, an effort of the "attention industrialist" Mark Zuckerberg. Yuan argues that the personalized digital learning that such platforms provide invariably figures students as possessive individuals who learn to valorize their own educational self-proprietorship. It is this same conception of education, as we see through his approach drawing on phenomenological anthropology, that operationalizes attention as an individual possession in a world of attention capture. In chapter 11, "Attention Is All You Need: Humans and Computers in the Time of Neural Networks," the anthropologist Nick Seaver considers the role of attention in the current era of machine learning, both as a feature of individual minds and as a collective capacity for deliberation. He shows that in its cultural expressions in our present technological conjuncture, attention might best be understood as a "key symbol" rather than simply as an objective mental process or capacity. In chapter 12, "Medium Focus," the art historian Joanna Fiduccia focuses her attention on materiality, or more precisely on the materiality of wood. Meditating on the example of Robert Morris's 1961 audio sculpture, *Box with the Sound of Its Own Making*, and on the art-historical context of this work's legacy and misinterpretations, Fiduccia reveals the fundamental respects in which "materiality determines attention." In this instance, that means, among other things, that the woodworker does not simply make a pure artifact out of a sort of prime matter but rather must attend to the wood's irreducible qualities, which lead back ultimately to the singular circumstances of its generation and growth.

With two chapters, part IV, "Endgame(s)," is the shortest but by no means only a coda. Here the authors approach attention as a phenomenon of experience at the margins or limits of distinctly human life, whether at the boundary between life and death or between healthy mental functioning and psychological trouble. In chapter 13, "Attention Fast, Attention Slow: Obsession, Compulsion, and Holding Close," the psychiatrist and historian Yael Geller brings her expertise to bear on Paul Virilio's classic investigation of the speed of contemporary life. She considers this phenomenon in relation to two familiar diagnostic categories—obsessive-compulsive disorder and attention deficit disorder—to answer the question, What happens when speed mediates attention? In the contemporary world, she argues, these psychological conditions stand as two gravitational poles by which a fairly precise measurement of the "disintegration" of attention may be taken. Finally, in chapter 14, "Units of Intensive Care: Poetic Attention and the Precarious Body," the literature scholar Lucy Alford relates the very personal story of her mother's coma in an intensive care unit in 2015. Alford is concerned with the unique expression of attention in the act of keeping vigil, which is hyperfocused but also diffuse and of indefinite temporal duration, uncertain in its precise end and imbued with profound meaning from within. From her account of this vigil, Alford moves into a reflection on the embodiment of attention and the particular significance of this condition for poetic language.

The works collected in these chapters come from several disciplines, operate at multiple registers, and approach their object from many directions. Yet, collectively, we believe, they form a unity. And most importantly, they reveal the unity of their shared topic of inquiry. Attention is one and several things at once, and it takes a concerted interdisciplinary effort of expert scholars in diverse fields to reveal this.

We have arranged our desks, and we have thought attention. You hold the result. Read on, attentively. Let us see if what we have been thinking can cross the space between us. (If a circle pops into your head, you know where the blame should fall!)

NOTES

1. The brief, schematic historical/historiographical sketch that follows draws on a range of sources, but it is worth explicitly citing a debt to Crary's *Suspensions of Perception: Attention, Spectacle, and Modern Culture* (Cambridge, MA: MIT Press, 2001).
2. But compare, for a compellingly contrastive story: Paul North, *The Problem of Distraction* (Stanford, CA: Stanford University Press, 2011).
3. For the "Attention Activism" of the Friends of Attention coalition, and their justice-oriented curricular project known as the "Attention Labs" consult: www.friendsofattention.net. In the summer of 2023, the Friends of Attention launched a new institution of teaching and learning that aims to center a movement of active resistance to the commodification of human attentional capacities: the Strother School of Radical Attention: www.attentionschool.org. Also relevant: D. Graham Burnett and Stevie Knauss, eds., *Twelve Theses on Attention* (Princeton, NJ: Princeton University Press, 2022).
4. In this regard, consider the work of ESTAR(SER), and its research on the so-called *Avis Tertia*, or "Order of the Third Bird."

I

HISTORIES OF ATTENTION

1

THE DISCOVERY OF ATTENTION

RICHARD J. SPIEGEL

A SCENE OF ATTENTION: EUREKA!

In February 1908, Edward Titchener, the British-born, German-trained, and American-based experimental psychologist, delivered a series of lectures at Columbia University. Titchener's appearance in Morningside Heights drew a healthy crowd. It was assured to be a good show. Not only was Titchener the most influential early experimental psychologist in the United States, but, befitting a man known for both intellectual and rhetorical verve, he was also a consummate showman. At Cornell, Titchener directed a small private orchestra. Reserving the "leading role for himself," he reveled in commanding the fear and esteem of his small ensemble. And, as his students keenly observed, the same pleasure he took in the theatrics of "showmanship" flowed directly into his performances at the lectern: "For serious student and famous psychologist that he was, he yet carried out his most ordinary pedagogic duties with all the pomp, the glamour, the fantastic unreality that called for opera bouffe music." He even took to the flamboyant habit of retiring his "baggy tweeds" and adorning himself "resplendent in the crimson of his Oxford gown" when lecturing in his laboratory. "It gives me the right to be dogmatic," he once brashly remarked to a student.[1]

Before his audience at Columbia, Titchener no doubt appareled himself more soberly. Relying on his oratory to carry the show, he conjured the interest of his audience by posing a question he considered of epochal significance, a question that cut to the heart of the entire intellectual project on which his research and reputation stood. "What, after all," he asked, pantomiming a layperson, "has the experimental method done for general psychology?" Titchener confessed that the question nakedly betrayed a gross ignorance. But its sheer simplicity could easily catch one off balance. For the influence of the exacting protocols of the laboratory on the study of the mind had been so total, he said, that the right answer was simply put: "Everything." The experimental method was responsible for all that one holds to be "right" and "irrefutable" in psychology. But he pressed on. Should we wish to be more specific, one could point to three things: "the complete recasting of the doctrine of memory and association, the creation of a scientific psychology in individual differences, and the discovery of attention." Of these three achievements, he homed in on the last, giving credit to Wilhelm Wundt, his own doctoral supervisor in Leipzig. Anticipating winces of suspicion from listeners who knew that neither the term nor a basic appreciation of the mental phenomena that it designated was of a recent vintage, Titchener quickly tightened his claim: "What I mean by the 'discovery' of attention is the explicit formulation of the problem; the recognition of its separate status and fundamental importance; the realization that the doctrine of attention is the nerve of the whole psychological system, and that as men judge of it, so shall they be judged before the general tribunal of psychology."[2] By so defining "discovery," Titchener wrangled his claim into an intermediate zone between histrionics and scholastic sobriety. With the boldness of his assertion and the caution of his qualification, he cogently braided together the affectation of high drama with the seriousness of truth. Doubtless, Titchener lived up to the hype.

Critically minded scholars spread across the histories of literature, philosophy, and science have since been keen to argue that attention was hardly the unique preoccupation of experimental psychologists. They show that attention was widely theorized and debated across the breadth of the Enlightenment.[3] How can experimental psychology have discovered attention if it had been so widely discussed in the eighteenth century, long before psychophysics had been a glimmer in Fechner's eye? By and large, such revisionist arguments elide Titchener's carefully circumscribed criterion of discovery, as if he had had no idea that philosophers from Alexander Baumgarten to Thomas Reid published sustained discussions on attention. Yet, interestingly, Titchener was not so aloof. In the years before his lecture, a spate of publications explored precisely the theorization of attention from the early Enlightenment onward.[4] Titchener didn't omit the counterevidence but only shuffled it offstage, including it in the endnotes of the printed version of his lectures. One might see this as a sleight of hand. Perhaps it is a sign that the temerarious rhetorician and the judicious scholar uneasily balanced within Titchener's person momentarily lost their footing. Perhaps. But the point was neither that attention hadn't been discussed at all before the emergence of experimental psychology, nor that it hadn't held import in thinking about cognition. It was about whether attention could be seen as the most important feature of the mind, "the nerve of the whole psychological system." Since attention is once again a subject of "fundamental importance" in the sciences of mind and brain, understanding when attention first became a matter of such central significance in modern psychology should be of renewed concern.⁵

In this chapter, I strengthen the argument against Titchener and those who subscribe to his periodization by showing the key importance of attention in eighteenth-century central European moral philosophy. I further marshal a categorical declaration of attention's significance to show just how central attention could be in late Enlightenment theorizations of the mind. Yet this example is instructive not only because it strongly refutes Titchener's point. It also helps us better grasp why, even if attention was a locus of intense concern in the eighteenth century, the fundamental importance of attention in the Enlightenment has long remained so obscure. In other words, if attention was as significant in the Enlightenment as Titchener's modern critics claim, then we also need an account of how his claim could have seemed at all plausible to his audience in Morningside Heights—just as it remains plausible for many today.⁶

ATTENTION AND AUTONOMY IN THE ENLIGHTENMENT

Christian Wolff's systematic philosophy exercised a tremendous influence on eighteenth-century intellectual life. With universities in crisis and a growing reading public, Wolff deliberately made his philosophy accessible to a broad audience. Coming from a modest background (his father was a tanner with some education), Wolff broke with tradition, indeed going even further than Christian Thomasius before him in lecturing and publishing his ideas in German, "so that," as he said, "even those who have not studied can obtain it." He believed strongly in the possibility of popular enlightenment and directed his efforts to creating a comprehensive philosophical system. The term for *enlightenment* in German (*Aufklärung*) referred not only to the optimism of a future-oriented era but also meant "clarification" or "demystification." To this end Wolff committed himself fully. Wolff tried to produce a coherent and integrated philosophical system, reaching from logic to cosmology, the principles of which relied in the first instance on empirical or "historical" experience.[7]

Wolff marked a turn in the early Enlightenment in Germany, a pivot in emphasis from natural law's concern with commands and rules to psychology's occupation with sentiment and passion as the source of morality. In his psychology, Wolff held that the soul was a simple immaterial substance. As perceptions, whether inner or outer, affected the soul, so the soul represented the world. Mental representations (*Vorstellungen*) contained propositions about the world marked by two qualities: clarity and distinctness. The first concerned the ability to differentiate one representation from another, the second the ability to differentiate the parts making up a complex representation. After Leibniz, Wolff saw everything in terms of perfection and imperfection. Things in the world that contain multiple parts are more perfect for the number of parts they contain and the simplicity of the principle organizing the harmonious functioning of those parts toward a particular end.

Wolff translated this thinking into his theory of desire and passion. He claimed that recognizing perfection gives us pleasure and that recognizing imperfection brings us pain. While the experiences of pleasure and pain correspond to what is good and evil, respectively, we may nevertheless experience pleasure and pain imprecisely and thus confuse one for the other. To act upon indistinct representations opened the possibility of acting against the essential will of attaining ever greater perfection.

It left one passive and in servitude to affective, irrational hungers. Only by a clear and distinct understanding of how acting upon a desire served not only one's pleasure but also the harmonious functioning of the social and cosmic whole could one act freely and rationally. The clearer and more distinct one's thoughts, the more perfect and freer one was. In the last instance, freedom was a binary decision: to refrain from or to act upon a desire based on the clarity and distinctness of a representation.[8]

At the most basic level, the perfection of reason relied on "attention," the primal "faculty" that governed the ability to make us "more conscious" of a given representation.[9] Wolff's discussion of attention reached a wide audience through his texts and those of his disciples. In the *Metaphysics*, which became a state-mandated textbook in Prussia, Wolff's student Alexander Baumgarten distilled Wolff's discussion of attention and its obverse, "abstraction," in his treatment of the intellect (§624–31). He noted how the resolving power of attention made a given phenomenon clearer while rendering the remaining phenomena in the field of consciousness more "abstract." It was because of this selectivity that various figures in the eighteenth century pointed to attention as a way of avoiding pain or blocking out sometimes cacophonous surroundings, typically while in the pursuit of a nobler task and often evoking strong resonances with neostoic philosophy.[10] Yet attending to the minute parts of any phenomenon was necessary for seeing the whole more distinctly. The movement of one's focus between parts and whole and the dynamism between attention and abstraction were the basis of comparison (*comparatio*) and reflection (*reflexio*). But even as part of a dynamism, attention remained paramount. It was the safeguard that regulated against flights of imaginative fancy.

By the middle of the eighteenth century, roughly coincident with Wolff's death in 1754, an ascendant philosophical program moved to displace the predominance of Wolff's "school philosophy," which had come to look like a new form of Scholasticism. The so-called popular philosophy of the high Enlightenment in Germany was characteristically a combination of Wolffian ideas and British empiricism, a common-sense philosophy with a humanistic emphasis on language and literature. In the popular philosophy, which muted metaphysical problems in favor of pragmatic philosophical instruction meant to groom a literate public to lead a virtuous life in service to an absolutist state, the attention–abstraction duality as the basis of rationality and autonomy was a mainstay. From Moses Mendelssohn to Johann Georg Feder, attention was a touchstone in theorizing freedom and reason. And, as such a superior category, attention was described by popular philosophers in magnificent ways. For Johann Sulzer, attention was "the light of the soul." For Georg Meier, it was the "source of all clarity of knowledge." For Karl Franz von Irwing, it was "the mother of all active ideas." For Dietrich Tiedemann, it was the "effluence of self-activity." Christian Garve, perhaps the foremost representative of popular philosophy, noted, "The most essential thing . . . [for] inner freedom of the mind [*Geistes*] . . . [is] the choice of the things to which [one] directs his attention." Garve described the centrality of the scrutinizing power of the mind as the "microscope of experience."[11]

PHILOSOPHY, PHYSIOLOGY, AND ANTHROPOLOGY

From the 1760s, the popular philosophy that predominated as the chief idiom of post-Wolffian philosophical discourse melded with a growing body of research in physiology and medicine.

A cohort of doctors, who fashioned themselves as "philosophical physicians," set out to drive empirical psychology and the "philosophical physiology" of Albrecht von Haller and his students into a single domain that could serve diagnostic and therapeutic functions in medicine. They aimed to develop the general "science of man," which would account for the whole person and be medically useful. By the late 1760s, they breathed new life into a centuries-old domain of knowledge that bound together theology, medicine, and philosophy, known as "anthropology."[12]

In 1772, the Saxon doctor Ernst Platner published his *Anthropology for Physicians and the Worldly*. Although Platner set himself only the relatively modest aim of giving doctors a stronger foundation in philosophy (since, as he noted, philosophers have a better knowledge of the body than physicians do of the soul), the text's tremendous success broadly defined the development of anthropology over the subsequent decades. Anthropology stood at the confluence of the soul and the body. The task of anthropology, he said, was "to consider together the reciprocal relationships, limitations, and connections [of] body and mind" (xvii). Like Haller, Platner had no qualms taking a dualistic metaphysics as axiomatic. The unity of consciousness relied on the simplicity of the spiritual substance. Since all matter was compound, the spiritual substance was immaterial. And while he stated that the object of his inquiry was to understand how movements in the material of the body produced ideas in the soul and vice versa, "it would betray the greatest ignorance," he held, "if one hoped to discover [the] secret [of the communion of the soul with the body]" (x).

Setting higher-order metaphysical questions aside, Platner's exposition drove as close to that great secret as possible. After an exhaustive discussion of how physical sensations communicated by the body's nervous system produced effects in the soul, he concluded that the movement of "nerve fluid" in the brain matter "is necessary for attention and consequently for mental representation, really effecting the soul." Yet he immediately clarified that impressions trafficked by nerve fluid are not "brought into the soul, as an object is depicted in the eye through light." His claim, he believed, was more circumspect: that the "movements of the life spirits in the brain . . . put the cognitive power [*Erkenntniskraft*] of the soul in motion." He continued, "When this happens, the soul directs its spiritual attention to the inner sensual impression. With this act the soul seems to work into the inner impression, and through the brain matter, and even through the nerves of the sensual instruments, back to the object" (§310–11). If, for Platner, the communion of the soul and body remained a secret, attention—the directed representational power of the soul that was activated by and, in turn, acted upon the nervous system—came about as close to the secret as possible. He even felt it necessary to clarify, "Through my system of the physical causes of attention, I have [not] wanted to invent the secret to explain the communion of the soul with the body" (§300). For Platner, attention was a sort of placeholder for the underlying problem of mind–body reciprocity.[13]

Despite its modest ambitions, Platner's tract was an instant success. It served as a model for the field of anthropology and became a touchstone in the creation of anthropology courses in universities.[14] At the recently founded Karlsschule outside Stuttgart, Platner's text had a strong impact on Jacob Friedrich Abel, a promising young philosopher now known principally for his influence on Schiller's early thought. Attention assumed a central position in Abel's work, showing the mark of Wolff and his followers, who positioned attention as the cognitive basis of moral autonomy, and Platner, who viewed it as the hinge between the material body and the immaterial soul. Schiller's earliest philosophical dissertations at the Karlsschule betray the same influences as those of his teacher, though Schiller, intellectually ambitious but anxious about the religious implications of materialist physiology, grasped tightly onto attention as the solution that would provide a minimum of agency to the soul. "The soul," Schiller said in his 1779 *Philosophy of Physiology*,

> has an active influence on the organ of thought. It can make the material ideas stronger and at discretion stick to them, and thus it also makes the spiritual ideas stronger. This is the work of attention. It has power over the strength of motivations [*Beweggründge*]; indeed it is [attention] itself that makes motivations. And now it would be rather decided what freedom is. . . . All human morality has its basis in attention, i.e., in the active influence of the soul on the material ideas in the organ of thought. . . . It is attention, then, through which we fantasize, through which we reflect, through which we separate and write, through which we will. It is the active influence of the soul on the organ of thought that accomplishes all of this. And so, the organ of thought is the true tribunal of the mind [*Verstand*]. . . . It is then completely dependent except for attention.

Discomfited by the implications of materialist physiology, Schiller turned to attention as the necessary solution to constructing a dualist theory that would secure the primacy and immateriality of the soul over the "*influxus physicus*" of the body. Although Schiller's examiners rejected his dissertation for, among other reasons, its "flowery style," which betrayed the "ferment of youth," Schiller's subsequent dissertation in 1780 reprised some of the key sentiments expressed about attention in the earlier text.[15]

Auspiciously, Schiller wrote his early dissertations only a few years before the publication of Kant's watershed first critique in

1781, marking by conventional periodization the transition from the high to the late Enlightenment in Germany. The publication of the transcendental philosophy shook not only Schiller but the whole German intellectual world. The Kantians' successful tirades against the popular philosophers, whom they lambasted for having neither the systematicity of Wolff nor the analytical perspicacity of Kant, galvanized factions over whether they were for or against Kantian philosophy. Even Platner, whose anthropology earned universal praise on publication, tried to reconcile himself with the Kantians in Jena for fear that his reputation would suffer otherwise.[16]

Yet, despite such ecumenism, Kant's critiques and their dogged champions drove a wedge between anthropology and philosophy that left a profound and lasting fracture. That fracture can be seen not only in the subsequent evolution of the disciplines but also, and perhaps most emblematically, in the sundering of Herder and Kant's once close personal and intellectual relationship. The idiom of eighteenth-century popular philosophy accommodated Kant's early philosophical commitments and provided the intellectual basis for his friendship with Herder throughout the 1760s. Yet, as Kant found his way to the critical philosophy, he and Herder drifted apart. Gradually, the two pursued diverging paths, with Kant trending decisively toward the dogmatic, anti-empirical, and ahistorical and Herder, holding on to his earlier commitments, trending toward the anthropological, empirical, and historical.[17] The bifurcation is evident in Kant's course on anthropology and his treatment of attention.

In his *Anthropology*, Kant followed the distinction between attention and abstraction from Baumgarten's *Metaphysics*. He treated both as a feature of faculty psychology. But he departed from the empirical psychology of popular philosophy and from the anthropology of Platner when he insisted that abstraction, "the *turning away from* an idea," was of much higher value than attention: "To be able to abstract from a representation, even when the senses force it on a person, is a far greater faculty than that of paying attention to a representation, because it demonstrates a freedom of the faculty of thought and the authority of the mind, in *having the object of one's representations under one's control (animus sui compos)*." Consequently, he concluded, "the faculty of *abstraction* is much more difficult than that of attention, but also more important, when it concerns sense representations." These passages not only betray Kant's distrust of sensory experience as the basis of autonomy but also show how he mapped that distrust onto the eighteenth-century attention–abstraction duality. Unlike for Platner, who viewed attention as a placeholder for understanding mind–body reciprocity while acknowledging that attention hardly resolved the underlying metaphysical issues at stake, Kant decisively shunted attention onto the side of the body, as a largely "involuntary" cognitive response of physical action and reaction. Consistent with the pragmatic orientation of his *Anthropology*, Kant illustrated his point in terms of practical mores and sociability: "Many human beings are unhappy because they cannot abstract. The suitor could make a good marriage if only he could overlook a wart on his beloved's face, or a gap between her teeth." The faculty of attention, in other words, was tantamount to unseemly fixations on sensory irregularities. It is only "prudent," he concluded, "to *look away from* the misfortunes of others," based on the "strength of mind" furnished by "the faculty of abstraction."[18]

By contrast, in his prize-winning essay on the origins of language, Herder followed Sulzer's commitment to attention as the source of reflective consciousness and deliberateness (*Besonnenheit*).

Applauding Sulzer's attempt to "clarify the metaphysics of the soul" from "*physical experiments,*" Herder submitted that "the first judgment of the soul" is the ability to fix the mind on an object. "The human being demonstrates reflection," he said,

> when the force of his soul operates so freely that in the whole ocean of sensations which floods the soul through all the senses it can, so to speak, separate off, stop, and pay attention to a single wave, and be conscious of its own attentiveness. The human being demonstrates reflection when, out of the whole hovering dream of images which proceed before his senses, he can collect himself into a moment of alertness, freely dwell on a single image, pay it clear, more leisurely heed, and separate off characteristic marks for the fact that this is that object and no other.[19]

Later, in his 1775 essay, "On the Cognition and Sensation of the Human Soul," he stated forthrightly, "The phenomenon of human freedom lies most deeply in the voluntary [*willkürlichen*] attention to pursue one side of the universe or to abstract from it."[20] For Herder, attention and abstraction remained two sides of the same coin, and both remained rooted in experience. Kant, however, pried them apart to ennoble the latter and denigrate the former, a firm indication of his commitment to differentiating himself from the popular philosophers of the eighteenth century. Since the critical philosophy insisted that the subject legislated itself in the spontaneous activity of reason—that the agent was morally *causa sui*—Kant rejected the notion that attention, with its strong association with what Wolff had called "historical" experience, could serve as the countercausal source of human freedom. Autonomy derived not from without but spontaneously from within.

ATTENTION AFTER KANT

At this point, the answer to our question should be coming into focus. If attention was so central in eighteenth-century philosophy of mind, how could Titchener's claim about the discovery of attention have appeared plausible to his audience in 1908? The answer might be just this simple: the vanquished do not enjoy the pleasure of writing their own history.

Kant's place in the history of philosophy is notable not only because of his titanic influence on the subsequent course of the history of ideas but also because, in presenting his philosophy as the dialectical supersession of Enlightenment antinomies, he deeply shaped how the history of philosophy would subsequently be told. According to Kant and his acolytes, the critical philosophy yoked together impossible opposites. It unified rationalism and empiricism, subjectivism and scientism by assimilating them into an account of the transcendental conditions of human subjectivity. As Ian Hunter insightfully remarks, "Considering that it was Kant himself who first viewed the history of philosophy in these terms, dialectical philosophical history is uniquely suited to demonstrating the epochal significance of Kantian philosophy, and may indeed be regarded as a subspecies of that philosophy."[21] Kant's and his students' attacks on the popular philosophy of the eighteenth century were not only decisive flash points that led scholars across the disciplines to take sides, but they were also subsequently elevated by the rising tide of dialectical philosophical history, relegating virtually all philosophy in Germany from the high Enlightenment into one of two categories: lesser derivations of Wolff or piecemeal anticipations of Kant.

The reputation of eighteenth-century popular philosophy suffered an immense defeat in the aftermath of the Kantian revolution. Dilthey's summary dismissal of popular philosophy at the turn twentieth century was a damning balance sheet: "the *Essais* of those days with their 'noble popularity' . . . are today an immense waste of paper; [they are] vast, shallow waters through which even the literary historian works his way only reluctantly."[22] In his influential *Grundriss der Geschichte der Philosophie*, which went through four editions by the century's end (1865–67, 1869, 1877, 1896), Johann Erdmann, a Hegelian and professor of philosophy at Halle, sneered at eighteenth-century popular philosophy. It failed to produce "an organic combination of the two tendencies [of realism and idealism], in which the opposition disappears in a higher unity," he said.[23] In 1888, just two years before Titchener stepped foot in Wundt's lab, a group of Anglophone scholars began the "Library of Philosophy," an ambitious publishing program meant to address the considerable dearth of work on the history of philosophy in English. The first text in the series, meant as a general primer, was a translation of Erdmann's history. The reviewer of the translation in *Mind* lauded the text for having made "the whole field of historical philosophy . . . accessible to the English-speaking student."[24]

The "all-unifying mill of dialectical philosophical history" illuminates those philosophies that are adapted to it and obscures those that are not.[25] Three now relatively obscure figures—Jakob Fries, Friedrich Beneke, and Johann Herbart—along with their students and sympathizers, signal that the currents and emphases embodied in the eighteenth century were hardly assimilated to the dominant trend of post-Kantian philosophies. In contrast to the rise of idealist rationalism and speculation, each of these three continued to place empiricism and psychology at the heart of his philosophies. And, significantly, each continued to place central emphasis on attention.[26] Much about these figures remains mysterious, not least of all how they accommodated themselves to the Kantian tradition when they discarded much of what we now consider to be its most essential contents.[27] Their continued obscurity to this day is due in no small measure to the fact that they were overshadowed by the meteoric rise of Hegelianism in *Vormärz* Prussia. Hegel's close relationship with the Prussian state ensured that his devotees enjoyed the most prominent university chairs and the most influential posts in the educational bureaucracy across the empire. It was even rumored that Hegel personally drafted the memorandum that dismissed Beneke from lecturing at the University of Berlin.[28] As one prominent historian of philosophy recently remarked, the recalcitrant invisibility of Fries, Beneke, and Herbart is undoubtedly in part a result of the Hegelians' success in writing them out of the history of philosophy.[29]

Yet a discursive erasure is hardly the same as an erasure in fact. And the prominence of Hegelianism in the history of philosophy still carries more than a whiff of Prussian supremacy. In the Kingdom of Saxony and in the Habsburg Empire, where transcendental idealism never set down such deep roots as in Prussia, Herbartianism and not Hegelianism became the dominant school of philosophy from the *Vormärz* period. In other words, how things looked differed depending on whether one was standing in Berlin, Leipzig, or Vienna. For Franz Exner, the first champion of Herbartianism in Austria, one of the major failures of the Hegelians was that they were unable to adequately address so fundamental a topic as attention.[30] And yet, by the end of the century, the disparagement of eighteenth-century philosophy had become so profound that the centrality of attention in Herbart's psychology came to look not like the continuity of an older tradition but like the germ of a new one, anticipating an achievement that would ultimately go to Wundt.[31]

CONCLUSION

In the same year that Titchener delivered his lectures at Morningside Heights, another influential early experimental psychologist, Hermann Ebbinghaus, published the now-oft-quoted phrase, "Psychology has a long past but only a short history." Both Titchener and Ebbinghaus enthusiastically paraded the standard of the new discipline, insisting that scientific psychology was properly a *pure discipline.* They went to pains to cleanly sever it from the centuries-old tradition that saw it subordinated under the umbrella of philosophy. No less for Ebbinghaus than for Titchener, the advancement of disciplinary psychology relied on cleansing it of unobservable and unmeasurable philosophical speculation. Scientific psychology, Titchener said, should confine itself "to the bare 'what' of things" and set aside that which is beyond its ken.[32] The notion that psychology was born anew at the bench drove a wedge between psychology's philosophical past and its future, a future to be bound and defined by what could be adequately addressed in the laboratory. Titchener's claim that the new psychology had discovered attention extended the same line of thinking from the history of the discipline to the objects of its inquiry.

Knowing how adroitly Titchener enticed his audience in Morningside Heights with his signature panache, it is tempting to say that he made so bold an assertion as the discovery of attention plausible by his superlative command of academic stagecraft. Notably, Titchener's own view of attention was that, as one of the "elementary" categories of psychology, it was not an activity of but a sensation in the mind. And, knowing his penchant for theatrics and his belief that attention was not something one actively did but something one passively experienced, we might speculate that his ostentatious showmanship, which by his own confession entitled him to a certain arrogance, was a cultivated practice in manipulating the minds of his listeners. And yet, even if Titchener did believe that his command over the attention of his audience could command their convictions, we should not confuse the power of his rhetoric for that of history. To do so would grant him altogether too much credit. For what even Titchener could not see was just how so audacious a claim as the discovery of attention had been made possible by a force far greater than himself, a force in which the antinomies of Enlightenment philosophy remained unresolved.

NOTES

Thanks to D. Graham Burnett and Justin E. H. Smith for encouraging me to write this essay and to Katja Guenther and Tony Grafton, who provided invaluable comments on earlier drafts. All faults are my own, as are translations unless otherwise noted.

1. Grace Adams, "Titchener at Cornell," *American Mercury* 24, no. 96 (1931): 440; Edwin G. Boring, *Psychologist at Large* (New York: Basic Books, 1961), 18.
2. Edward B. Titchener, *Lectures on the Elementary Psychology of Feeling and Attention* (New York: Macmillan, 1908), 171–73.
3. E.g., Gary Hatfield, "Attention in Early Scientific Psychology," *IRCS Technical Reports Series* 144 (1995): 1–35; Michael Hagner, "Towards a History of Attention in Culture and Science," *MLN* 118 (2003): 670–87; Margaret Koehler, *Poetry of Attention in the Eighteenth Century* (New York: Palgrave Macmillan, 2012), 15–60.
4. E.g., Lemon L. Uhl, *Attention: A Historical Summary of the Discussions Concerning the Subject* (Baltimore, MD: Johns Hopkins University Press, 1890); David Braunschweiger, *Die Lehre von Aufmerksamkeit in der Psychologie des 18. Jahrhunderts* (Leipzig: Hermann Haacke, 1899); Max Dessoir, *Geschichte der neueren deutschen Psychologie*, 2nd ed. (1894; repr., Berlin: Carl Duncker, 1902), 417–20.
5. Owing to the rise of behaviorism and Gestalt psychology, research on attention experienced a precipitous decline about a decade after

Titchener's lecture. It was only with the rise of cognitive psychology after World War II that attention again moved to center stage. See Addie Johnson and Robert W. Proctor, *Attention: Theory and Practice* (Thousand Oaks, CA: Sage, 2004), 3.

6. The most influential work of humanistic scholarship on the history of attention remains Jonathan Crary, *Suspensions of Perception: Attention, Spectacle, and Modern Culture* (Cambridge, MA: MIT Press, 1999). Crary's influential Foucauldian argument, although dissimilar to Titchener's thinking in crucial ways, follows Titchener's periodization on the basis that one would be hard pressed "to find before 1850 an unconditional statement" insisting on the centrality of attention in psychology (21). Yves Citton's recent monograph indicates that Crary's text remains the signal work in the field, see *The Ecology of Attention*, trans. Barnaby Norman (Cambridge: Polity, 2017), 14–15.

7. My account of Wolff's moral psychology relies on J. B. Schneewind, *The Invention of Autonomy: A History of Modern Moral Philosophy* (Cambridge: Cambridge University Press, 1998), 431–37, quote on 434.

8. Christian Wolff, *Vernünftige Gedanken von Gott, der Welt und der Seele des Menschen* (Halle, 1720), esp. §439, 514, 519.

9. Wolff, *Vernünftige Gedanken von Gott*, §268–71, esp. §865–66. The best contemporary discussion of attention in Wolff's philosophy is Hans Adler, "Bändigung des (Un)Möglichen: Die Ambivalente Beziehung zwischen Aufmerksamkeit und Aufklärung," in *Reiz, Imagination, Aufmerksamkeit: Erregung und Steuerung von Einbildungskraft im klassichen Zeitalter (1680–1830)*, ed. Jörn Steigerwald and Daniela Watzke (Würzburg: Königshausen & Neumann, 2003), 41–54.

10. This feature of attention dovetailed neatly with the notion of patience in neostoic philosophies. E.g., Johann F. Zuckert, *Von den Leidenschaften* (Berlin, 1774), §66–68, 76, 81–84; Christian Garve, "Ueber die Geduld," in *Christian Garves Sämmtliche Werke* (Breslau, 1801), 1:45–46, 85–86.

11. Daniel O. Dahlstrom, ed. *Moses Mendelssohn: Philosophical Writings* (Cambridge: Cambridge University Press, 1997), 18–19; Johann G. H. Feder, *Logik und Metaphysik* (Vienna: Johann Thomas Edl. Von Trattern, 1783), 39–40; Sulzer, Meier, and Irwing quoted in Braunschweiger, *Aufmerksamkeit*, 66; Dietrich Tiedemann, *Handbuch der Psychologie zum Gebrauche bei Vorlesungen* (Leipzig: Barth, 1804), §70 on 121; Christian Garve, "Ueber die Muße," in *Anthologie aus den sämmtlichen Werken*

von Christian Garve (Hildburghausen: 1844), 141; Johan van der Zande, "The Microscope of Experience: Christian Garve's Translation of Cicero's 'De Officiis' (1783)," *Journal of the History of Ideas* 59 (1998): 90.

12. The idea of anthropology as a general science of the human encompassing its spiritual and physical aspects traces its roots to Reformation-era confessional polemics. The early modern and Enlightenment notions of anthropology are genealogically contiguous with but conceptually and methodologically distinct from the modern sense of the term, which designates the disciplinary study of human culture and society. On the Renaissance origins of the idea of anthropology, see Tricia M. Ross, "Anthropologia: An (Almost) Forgotten Early Modern History," *Journal of the History of Ideas* 79 (2018): 1–22.

13. On Platner, see Alexander Košenina, *Ernst Platners Anthropologie und Philosophie: der philosophische Arzt und seine Wirkung auf Johann Karl Wezel und Jean Paul* (Würzburg: Königshausen & Neumann, 1989).

14. Mareta Linden, *Untersuchungen zum Anthropologiebegriff des 18. Jahrhunderts* (Bern: Lang, 1976).

15. Abel placed attention centrally in both his moral philosophy and his physiological psychology. See Jacob Friedrich Abel, *Eine Quellenedition zum Philosophieunterricht an der Stuttgarter Karlsschule (1773–1782)*, ed. Wolfgang Riedel (Würzburg: Königshausen & Neumann, 1995), 110, 130, 221–236, 421–27, 439, 542, 572–74, 590; Friedrich Schiller, *Schillers Werke: Nationalausgabe*, ed. Benno von Wiese and Helmut Koopmann, (Weimar: Hermann Böhlaus Nachfolger, 1962), 20:26–28, 46; Wolfgang Riedel, *Die Anthropologie des jungen Schiller: Zur Ideengeschichte der medizinischen Schriften und der 'Philosophischen Briefen'* (Würzburg: Königshausen & Neumann, 1985). Examiners' reports quoted in Kenneth Dewhurst and Nigel Reeves, *Friedrich Schiller: Medicine, Psychology, and Literature* (Berkeley: University of California Press, 1978), 167–68.

16. Košenina, *Ernst Platners Anthropologie*, 19.

17. John H. Zammito, *Kant, Herder, and the Birth of Anthropology* (Chicago: University of Chicago Press, 2002).

18. Immanuel Kant, *Anthropology from a Pragmatic Point of View*, trans. and ed. Robert B. Louden (Cambridge: Cambridge University Press, 2006), 19–20 (emphases in original). In the first critique, Kant discussed attention only briefly in a footnote added to the second edition. On

the relationship between Kant's anthropology and that footnote, see Rodolphe Gasché, "One Seeing Away: Attention and Abstraction in Kant," *Centennial Review* 8 (2008): 1–28.
19. Johann Gottfried von Herder, *Philosophical Writings*, trans. and ed. Michael N. Forster (Cambridge: Cambridge University Press, 2002), 87–88, 88n.m (emphasis in original).
20. Quoted in Braunschweiger, *Aufmerksamkeit*, 66.
21. Ian Hunter, *Rival Enlightenments: Civil and Metaphysical Philosophy in Early Modern Germany* (Cambridge: Cambridge University Press, 2004), x.
22. Wilhelm Dilthey, "Freidrich der Große und die deutsche Aufklärung," in *Wilhem Dilthey Gesammelte Schriften* (Stuttgart: Teubner, 1992 [1927]), 3:174.
23. Johann Erdmann, *A History of Philosophy*, 3rd English ed., trans. Williston S. Hough (London: Swan Sonnenschein, 1892 [1890] 2: §293.9, on 310.
24. N.a., "Notes," *Mind* 13 (1888): 317–18; N.a., "A History of Philosophy," *Mind* 15 (1890): 132–33.
25. I borrow this wonderful turn of phrase from Hunter, *Rival Enlightenments*, x.
26. Dessoir, *Geschichte*, 194–203.
27. Frederick Beiser painfully tries to square this circle in his brief article, "Herbart's Monadology," *British Journal for the History of Philosophy* 23 (2015): 1056–73. Fries claimed that Platner and Kant were his greatest influences.
28. Klaus C. Köhnke, *Entstehung und Aufstieg des Neukantianismus: die deutsche Universitätsphilosophie zwischen Idealismus und Positivismus* (Frankfurt: Suhrkamp, 1986), 46–47.
29. Frederick C. Beiser, *The Genesis of Neo-Kantianism: 1796–1880* (Oxford: Oxford University Press, 2014), 11–12.
30. Deborah R. Coen, *Vienna in the Age of Uncertainty: Science, Liberalism, and Private Life* (Chicago: University of Chicago Press, 2007), 50.
31. Harry E. Kohn, *Zur Theorie der Aufmerksamkeit* (Halle: Ehrhardt Karras, 1894), 1.
32. Quoted in John J. Cerullo, "E. G. Boring: Reflections on a Discipline Builder," *American Journal of Psychology* 101 (1988): 563.

2

ATTENTION AND BOREDOM IN EARLY AMERICAN PSYCHOLOGY

HENRY M. COWLES

A SCENE OF ATTENTION: TEDIUM AND THE CATS

The graduate student is getting bored. With a chronograph in one hand and a clipboard in the other, he waits for a cat to escape a cage. Unbeknownst to the cat, there's a trick to it: pushing a lever, pulling at some string. Pawing and pawing, the cat will eventually, accidentally lift the latch. Depending on the cage's configuration (and the cat's desperation), it could take a while. If and when it gains its freedom, the cat goes right back in. One escape doesn't mean much, after all: only in aggregate does the graduate student's subject—the learning process—come into view. This makes work a matter of waiting out all the trials and errors, fending off the boredom that inevitably arises. Some cats never learn, and those that do require dozens of repetitions. As the seconds turn to minutes and the ledgers of frustration fill up, our student fights to pay attention. Glancing from cat to clipboard and back again, he waits for the fateful move that spells (temporary) freedom for them both.

So, the student focuses. A growing sense of how a cat learns to escape leads him to track its eyes in particular—to attend to its attention. This is less about his subject and more about the student's self-interest: cats that watch their paws learn quicker. They are less boring. As the light streams in through the attic window and the clock ticks toward the end of his day, the student staves off his boredom by attempting to bore the cat. That's what success looks like: as a cat learns its lesson, the task becomes rote and the mind wanders—or seems to. There's really no telling, the student is realizing, what it's like to be a cat. Maybe it's his own boredom, his own attention that he's reading into the cat's behavior, an anthropomorphic recognition of repeating a well-learned task. Looking where the cat is looking, the student seeks the ebb and flow of boredom because he himself is feeling bored; he projects his attention *and* his affect onto the animal.

This is a scene from the "new psychology," a movement at the turn of the twentieth century through which German experimental procedures were adapted to American contexts. The student, Edward Thorndike, spent the 1890s scrutinizing a menagerie of cats, dogs, and chickens locked in cages of his own design. Both Thorndike's advisers—William James at Harvard and James McKeen Cattell at Columbia—played roles in this new psychology, albeit quite different ones. What made it new was not so much what these psychologists studied but *how* they studied it: the instruments they carried and the affect that accompanied their use. The chronograph and the clipboard remade psychology in the image of laboratory physics, reorienting the study of subjective experience around quantitative measures such as intensity and temporality. This transformation focused attention on mental states—like attention itself—that were capable of being measured. Such studies were, for many, boring. But rather than the study of attention making psychology boring, it was the boredom of (some) new psychologists that made attention visible to them as they took notes on cats, dogs, and one another.[1]

PSYCHOLOGY GETS BORING

Boredom has many dimensions: subjective and objective, affective and existential. Whereas some scholars link boredom to longer histories of malaise or ennui, others insist it is peculiarly modern—a state that flowered, with the use of the term itself, in the nineteenth century. In her genealogy of boredom, Elizabeth Goodstein shows how, "in the course of the nineteenth century, the traditional understanding of ennui as a malaise of longing is psychologized." The result, she argues, is that boredom became "an experience of time understood as a Newtonian, mechanical process." Extending her analysis, we can see the new psychologists, with their ticking clocks and tapping pens, picking up where Gustave Flaubert left off. Boredom becomes both an object of research and the defining affect of *doing* that research; it inhabits the space between attention and habit, between absorption in a new task and the freedom of a familiar one. This was the space claimed by the new psychology: as an object, boredom helped define the boundaries of consciousness; as an experience, it defined science itself.[2]

Boredom had enemies—including Thorndike's Harvard adviser William James. Having founded a laboratory and taken Thorndike's animals into his own home, James would seem to have championed the new psychology. But he hated it. The way this new generation pursued alien, even alienating procedures amounted to a "siege," fought through "patience, starving out, and harassing to death." Students like Thorndike were "spying and scraping, their deadly tenacity and almost diabolic cunning" fast becoming a requirement. James was horrified: "This method taxes patience to the utmost, and could hardly have arisen in a country whose natives could be *bored*. Such Germans as Weber, Fechner, Vierordt, and Wundt obviously cannot; and their success has brought into the field an array of younger experimental psychologists, bent on studying the *elements* of the mental life, dissecting them out from the gross results in which they are embedded, and as far as possible reducing them to quantitative scales." Boredom, on this reading, was not a regrettable result of important work; it *was* the work, transforming psychology from philosophical reflection to self-flagellation and making science hard to swallow.[3]

Part of the problem was how single-minded the new approach was. For James, addressing any problem worth studying required constantly altering the angle from which it was viewed. "Only in pathological states," he wrote in 1890, "will a fixed and ever monotonously recurring idea possess the mind." He seems to have had his colleagues in mind. James went so far as to invert an old adage about geniuses to prove his point: "*It is their genius making them attentive, not their attention making geniuses of them.*" Those who merely paid attention were the problem with psychology, not its solution. "It is probable that genius tends actually to prevent a man from acquiring habits of voluntary attention, and that moderate intellectual endowments are the soil in which we may best expect, here as elsewhere, the virtues of the will, strictly so called, to thrive." This attack was subtler than the earlier one but no less barbed: thumping away at a single experiment amounted to a sign of mental illness.[4]

Eventually, James handed off his laboratory to Hugo Münsterberg (a German, incidentally), whose catalog of laboratory apparatuses captured the material aesthetic of the new psychology (figure 2.1). Table after table of instruments, each dedicated to the harassing dissection James bemoaned. As a new generation of students remade the field with such tools, apostates like James turned their attention to other things. No longer tied to the laboratory, in practice or even in pedagogy,

FIGURE 2.1 Hugo Münsterberg's catalog of laboratory apparatuses

James attended séances, studied the history of religion, and dabbled in self-help. We can see this turn away from instrumental reduction in everything from his discussion of "the stream of thought"—irreducible as it was to measurement—to his defense of acting *rather* than waiting in "The Will to Believe." Why gather more and more data at a smaller and smaller scale when we can break the spell of boredom by asking new questions in new contexts?[5]

Psychology was not alone in its boredom. Many fields were changing in the decades around 1900, as science's public image and self-presentation narrowed into white coats and dry rhetoric. This was the heyday of what Lorraine Daston and Peter Galison call "mechanical objectivity"; according to Theodore Porter, this was when science became technical. The standardization, if not to say *industrialization*, of everything from how research was conducted to how findings were published remade science in the image of the assembly line. There were exceptions, to be sure, and opponents aplenty. But the watchword of efficiency was everywhere, from scientific management to self-improvement, and the effect was uniform: subdivided tasks pursued serially in the name of a larger (if largely invisible) whole. Science, like much else, was becoming quantitative, stepwise. Indeed, one could say this was the moment and the manner in which science became modern. Success across a range of fields was becoming a matter of sustaining attention, of ascetically tolerating tedium.[6]

ATTENDING TO ATTENTION

As the new psychologists fanned out across the United States, founding departments and laboratories to anchor their approach, an experimental vernacular emerged. For both champions and opponents alike, this centered on the study of so-called reaction times: the gaps between sense impressions and motor responses. Reaction times could be observed both subjectively and objectively, with researchers recording one another's with watches. Perhaps the most common setup was to have one subject press a key upon hearing a sound while the other timed the gap between the sound and the action. More important than the mode of measurement was the *fact* of measurement, or rather of measurability. Because reaction times could be recorded down to the millisecond, they accorded with the values of precision sweeping the field. Made famous by Wilhelm Wundt and his colleagues at Leipzig, the reaction time became a building block for the new science of the mind, a window into the smallest unit of experience.[7]

Attention was crucial to such studies—not as a *dependent* variable to be analyzed but as an *independent* variable to be manipulated. For example, subjects asked to focus their attention on the key to be pressed had different reaction times from those who focused on the sound they expected to hear. Psychologists ended up disagreeing over the nature of these changes—and thus, over the reaction time itself. On one side were those like Wundt and his student Edward Titchener, who argued that reaction times were the same for everyone. On the other side were psychologists like James Mark Baldwin, who saw different "types" of *reactors*—differences that emerged when subjects were asked to *attend* in specific ways. Titchener saw Baldwin's "types" as artifacts of experimental aptitude, or *Anlage*—it was their experience doing experiments, not their fundamental attention, that differed. Baldwin, in turn, accused Titchener of using the idea of *Anlage* to throw out results he didn't like. Despite their differences, the sides shared an approach to attention: it was something they *used*, not something they *studied*. In the end, their dispute was less about reaction times and more about attention—specifically, the *uses* of attention, reframed as an indispensable tool for the study of minds, the *sine qua non* of science itself.[8]

But attention soon slipped from independent to dependent variable. As practitioners poked and prodded their laboratory subjects, they gradually realized that William James's claim that "Every one knows what attention is" was less true than it seemed. The so-called Chicago School, for example, began to see attention less as an individual ability (Titchener's *Anlage*) or a series of types and more as a spectrum across which we all move. This new, expansive sense of attention emerged in a famous study on the effect of practice on reaction time (figure 2.2).

In place of Baldwin's taxonomy of reactors (here dubbed "reagents"), the Chicago team unearthed an *ecology* of attention, a landscape to be explored using reaction-time studies. In other

Table Showing the Results of Practice

Sen. Organ.	Mot. Organ.	Focus of Atten.	Reagents								
			A			M			J		
			No.	Time in σ		No.	Time in σ		No.	Time in σ	
				First Third	Last Third		First Third	Last Third		First Third	Last Third
Ear	Hand	Sen.	560	195	133	420	163	132	160	185	173
		Mot.	540	149	127	380	178	134	165	169	159
	Foot	Sen.	145	182	168	220	138	133	125	218	208
		Mot.	155	159	150	230	145	134	115	204	196
	Lips	Sen.	130	132	122	125	117	108	160	169	155
		Mot.	120	125	116	125	112	106	140	157	146
Eye	Hand	Sen.	100	206	173	100	153	125	100	180	173
		Mot.	100	193	150	100	176	130	100	193	168
	Foot	Sen.	100	218	170	110	160	153	125	229	183
		Mot.	100	170	151	115	153	148	125	199	175
	Lips	Sen.	130	141	135	125	144	138	100	193	179
		Mot.	120	133	127	125	136	133	100	166	165

FIGURE 2.2 The effects of practice on reaction time

words, they inverted the priority of reaction time and attention, now using the former to study the latter rather than the other way around.⁹

This transformation was capped off by one of the period's most famous papers: John Dewey's "The Reflex Arc Concept in Psychology." Dewey, the leader of the Chicago team that was turning attention into a dependent variable, set out to sharpen their work into an attack on reaction-time studies in general. The very idea, he argued, was founded on a false distinction between stimulus and response. While the two *seem* distinct in everyday life—we hear a noise and flinch, we see a friend and smile—the distinction falls apart under scrutiny. "At one time," Dewey writes, "fixing attention, holding the eye fixed, upon the seeing and thus bringing out a certain quale of light is the response, because that is the particular act called for just then; at another time, the movement of the arm away from the light is the response. There is nothing in itself which may be labelled response." We act and react, or think and behave, not in a series of arcs from one place to another, but rather in circuits of coordinating behaviors that blend into one another as they produce more complex mental states—like attention.[10]

The Chicago School's inversion of attention and reaction affirmed James's view that the new psychology's boring methods had thrown the baby of experience out with the bathwater of introspection. Our reactions always depend on what captures our attention. As Dewey put it: "If one is reading a book, if one is hunting, if one is watching in a dark place on a lonely night, if one is performing a chemical experiment, in each case, the noise has a very different psychical value; it is a different experience. In any case, what proceeds the 'stimulus' is a whole act, a sensorimotor coordination." These were the stakes that James felt psychology was taking for granted. In elevating reaction times, the field missed what it took to read a book or stare off into space: attention. The fact that science itself now required such powers of attention was a sign of just how perverse it was that psychologists took this capacity for granted.[11]

WHAT IS IT LIKE TO BORE A CAT?

The transformation of attention from independent to dependent variable was neither completely novel in the work of the Chicago School nor completely dominant thereafter. But we do see a shift at the turn of the twentieth century from reaction time to attention, a shift that accompanied (and to a certain extent bolstered) an interest in attention beyond the academy: among teachers eager to corral their students and advertisers keen to keep customers. Inside and outside the laboratory, attention was both a subjective state and a desirable commodity, a necessary tool for navigating a complex world and a monetizable asset central to a burgeoning attention economy. This dual existence was mirrored in the ambitions of attention researchers themselves. While many remained focused on identifying and laying the building blocks of a scientific theory of cognition, others sought to make attention pay in the wider world—to position themselves as experts in the identification and coercion of attention practices beyond the laboratory. Some psychologists did both at once.[12]

One such practitioner was Thorndike's Columbia advisor, James McKeen Cattell. Cattell, who had been a student of Wundt, appears throughout the reaction-time debates of the 1890s because of the studies he conducted and published during his time in Leipzig. Some of his papers contradicted the theory of simple reaction put forward by Titchener, who rejected them as artifacts of the particular subject involved—in this case, Cattell. "Professor Cattell himself is so exceedingly practised a reagent," Titchener wrote, "that one may suspect automatism in his case." What Titchener meant was that Cattell was no longer typical—he had turned attention into a kind of habit. The line between attention and habit was being probed by others in this period, including Gertrude Stein in her famous "automatic writing" experiments under the loose direction of James at Harvard. Just as it had in Chicago, attention was shifting from independent to dependent variable, inserting itself where it was not wanted and preventing Stein and her collaborator from homing in on the automatism they sought. "Our trouble," they wrote, "never came from a *failure of reaction*, but from a *functioning of the attention*. It was our inability to take our minds off of the experiment that interfered." For Stein, as for Cattell, the experience of experiment and the inability to be bored or distracted forced attention into the spotlight.[13]

Cattell's students sought to imitate their mentor's automatism even as they began to adapt his mental tests for the study of nonhuman animals. Thorndike was one such student, having been lured away from Harvard for a finishing fellowship at Columbia in 1897. The challenge of adapting Cattell's methods to cats was finding a way to move beyond introspective self-report to exclusively behavioral studies. Since you could not simply ask cats to pay attention (or ask them to reflect on whether their attention had been sustained), Thorndike had to design experiments that were simple enough for cats to complete but interesting enough so they paid attention to the task at paw. Thorndike found a path forward in a study of bugs by the British polymath John Lubbock: "In order to test their intelligence, it has always seemed to me that there was no better way than to ascertain some object which they would clearly desire, and then to interpose some obstacle which a little ingenuity would enable them to overcome." Hunger, in other words, was how you could tap into any animal's mind: "Never," Thorndike wrote, "will you get a better psychological subject than a hungry cat."[14]

Initially, Thorndike assumed that it was attention *to the desired object* that was necessary, hence his leaving food "outside in sight" of the cat. The idea was that attention to the food would both motivate attempts to escape *and* guide the process of achieving it. In reality, hunger and attention could work at cross purposes. If the cat was *too* eager, it paid less attention—which, Thorndike realized, was what really mattered. Attention was "often correlated with a lack of vigor," with the result that an "absence of a fury of activity let him be more conscious of what he did do." Like Angell and Moore, Thorndike began elevating attention. "The really effective part of animal consciousness, then, as of human," he concluded, "is the part which is attended to; attention is the ruler of animal as well as human mind." But this finding did not mean that attention was the same for all animals. The "cat watching me for signs of my walking to the cage with fish is not in the condition of the man watching a ball game, but in that of the player watching the ball speeding toward him." Attention, in making the transit from tool to target, was being broken apart, complexified.[15]

The same was true of Thorndike's attention. It is the *scientist's* attention, strained by boredom, at the heart of the scene with which we began. A sense of that strain comes through in the data (figure 2.3).

Beyond all the waiting and the timing and the feeding, Thorndike manually guided the animal through the necessary

Individual	Apparatus	Time in which impulsive activity failed to lead to the act	Number of times the animal was put through the movement	Time in which this experience failed to lead to the act	Time of final trial
Cat 1	F (String outside unfastened)	55.00	77	120.00	20.00
Cat 5	G (Thumb-latch) - -	57.00	59	55.00	10.00
Cat 7	G (Thumb-latch) - -	50.00	30	35.00	10.00
Cat 2	G (Thumb-latch) - -	54.00	141	110.00	20.00
Dog 2	BB1 (0 at back, high) -	48.00	30	80.00	60.00
Dog 3	BB1 (0 at back, high) -	20.00	85	55·00	10·00
Dog 2	M (Lever outside) - -	15.00	95	140·00	30·00
Dog 1	FF [1] - - - - -	30.00	110	135.00	60.00
Chick 89	X (See page 53)	20.00	30	60.00	30.00
Cat 13	KKK [2,3] - - - -	40.00	65	60.00	10.00

FIGURE 2.3 Thorndike's data on attention

movement over, and over, and over. Small wonder that the first topic he took up after completing his dissertation was mental fatigue—a feeling that consisted, he argued, "*not in the amount done, but in the increased number of mistakes or inappropriate reactions.*" All those studies of attention, if pursued for too long, led to lapses in attention—and in turn to mental fatigue and error. In making attention the object of his attention, Thorndike learned how boring it was to bore a cat.[16]

CONCLUSION

The first historian of the new psychology had an inauspicious surname. Edwin Boring trained with Titchener at Cornell and went on to publish his iconic *A History of Experimental Psychology* in 1929. The book is biographical and—as readers have been quick to point out—boring. The biographical focus, at least, was intentional: the book credited the new psychology to the particular personalities of its champions. This raised an interesting question: "If personalities lie, in part, back of psychology, what lies back of the personalities?" Boring's answer was similar to James's: the ability to sustain attention in the face of crushing boredom. Gustav Fechner, to whom both men attributed the field's modernization, took it so far that he "developed, as James diagnosed the disease, a 'habit-neurosis'" that left him bedridden for years. Boring identified a similar tendency in Wundt, whose youth "seems to have been almost entirely a life of study, with few friendships and social activities." Among other things, "the lack of normal outlets for boyish energy fixed habits of study and thought upon Wundt that in part account for his later great accomplishment." For James, this tolerance for boredom was depressing; for Boring, it was essential—not just to psychology but to science.[17]

Ultimately, boredom won the day. The triumph of statistics over stimulation altered the content and context of psychological work. Laboratories, once crowded with apparatuses and animals, were emptied out. Take two scenes from an illustrated guide to the new psychology as evidence. According to Edward Scripture, the guide's author, "the struggle for ever increasing accuracy is the vital principle of all the sciences." It doesn't take much empathy to understand the affective results of moving from the social scene on the left of figure 2.4 to the "isolated room" on the right.

Isolation was intended to pinpoint mental elements by reducing all inputs, sensory or otherwise. Only then could psychologists be sure what they were studying. Boredom became a tool, if not a rule, essential to preserving not only objectivity but accuracy.[18]

FIGURE 2.4 An older laboratory and a newer laboratory (the "isolated room")

FIGURE 2.4 (*cont*)

This vision of psychology was more than boring—it was claustrophobic. The lengths to which Scripture went to stamp out error morphed into sensory deprivation in the pursuit of truth: "Of course, the experimenter, the recording apparatus and the stimulating apparatus are in a part of the building distant from the person experimented upon. He sits in the reacting room perfectly alone, knowing nothing of what is going on. The warning click of a sounder tells him to concentrate his attention; a click occurs in the telephone, or a Geissler tube flashes out, or an electric shock pricks the skin; he reacts in response and all is again quiet. All light and moving objects are, of course, excluded."

The student in the isolated room was not unlike a cat in one of Thorndike's cages. Both were "dissected," as James had put it, "out from the gross results in which they are embedded," their boredom—if not misery—reconceived as a rite of passage. Topics of study mattered less than the pain of studying. As Boring later reflected, "We are no longer so sure that it was really 'attention' that was being investigated, but all we need to accomplish here is to picture the enthusiasm of the time." People were excited; never mind about *what*. Even for those obsessed with attention, researchers' ability to *sustain* it mattered more than their claims *about* it. Science had been reconceived as strenuous, aptitude as the ability to attend. Boring was right.[19]

NOTES

1. Details of Thorndike's life can be found in Geraldine M. Jonçich, *The Sane Positivist: A Biography of Edward L. Thorndike* (Middletown, CT: Wesleyan University Press, 1968). The canonical account of this history was long Edwin Garrigues Boring, *A History of Experimental Psychology* (New York: Century, 1929). For a more recent account, see Kurt Danziger, *Constructing the Subject: Historical Origins of Psychological Research* (Cambridge: Cambridge University Press, 1990).
2. Elizabeth S. Goodstein, *Experience Without Qualities: Boredom and Modernity* (Stanford, CA: Stanford University Press, 2005), 112. Boredom is a subject of increasing interest in a range of fields. For a recent psychological overview, see James Danckert and John D. Eastwood, *Out of My Skull: The Psychology of Boredom* (Cambridge, MA: Harvard University Press, 2020). On its economic impact, see Sydney Ember, "The Boredom Economy," *New York Times*, February 20, 2021, https://

www.nytimes.com/2021/02/20/business/gamestop-investing-economy.html.

3. William James, *The Principles of Psychology*, vol. 1 (New York: Henry Holt, 1890), 185, 192–93. On James's complex relationship with the new psychology, see David E. Leary, "Telling Likely Stories: The Rhetoric of the New Psychology, 1880–1920," *Journal of the History of the Behavioral Sciences* 23, no. 4 (1987): 315–31. On his and others' ambivalence about introspection, see Deborah J. Coon, "Standardizing the Subject: Experimental Psychologists, Introspection, and the Quest for a Technoscientific Ideal," *Technology and Culture* 34, no. 4 (1993): 757–83. On quantification in the field, see Ruth Benschop and Douwe Draaisma, "In Pursuit of Precision: The Calibration of Minds and Machines in Late Nineteenth-Century Psychology," *Annals of Science* 57 (2000): 1–25.

4. James, *Principles*, vol. 1, 423–24. James explored genius at greater length in William James, "Great Men, Great Thoughts and the Environment," *Atlantic Monthly* 46, no. 276 (1880): 441–59. Such discipline had come to define methods across a range of scientific fields. See, e.g., Simon Schaffer, "Astronomers Mark Time: Discipline and the Personal Equation," *Science in Context* 2, no. 1 (1988): 115–45.

5. Hugo Münsterberg, *Psychological Laboratory of Harvard University* (Cambridge, MA: Harvard University Press, 1893). The plate appears between 12 and 13. For more on Münsterberg, see Jeremy T. Blatter, "The Psychotechnics of Everyday Life: Hugo Münsterberg and the Politics of Applied Psychology, 1887–1917" (PhD diss., Harvard University, 2014). On James's boundary work, see Francesca Bordogna, *William James at the Boundaries: Philosophy, Science, and the Geography of Knowledge* (Chicago: University of Chicago Press, 2008).

6. On mechanical objectivity, see Lorraine Daston and Peter Galison, *Objectivity* (New York: Zone, 2007), 115–90. On its relationship to attention, see 240–46. For Porter's account, see Theodore M. Porter, "How Science Became Technical," *Isis* 100, no. 2 (2009): 292–309.

7. On Wundt's reaction-time studies, see David Kent Robinson, "Reaction-Time Experiments in Wundt's Institute and Beyond," in *Wilhelm Wundt in History: The Making of a Scientific Psychology*, ed. R. W. Rieber and David Kent Robinson (New York: Kluwer Academic, 2001), 161–204.

8. For Titchener's version of the *Anlage*, see E. B. Titchener, "Simple Reactions," *Mind* 4, no. 13 (1895): 74–81. For the type theory, see J. Mark Baldwin, "Types of Reaction," *Psychological Review* 2, no. 3 (1895): 259–73. On their debate, see David L. Krantz, "The Baldwin-Titchener Controversy: A Case Study in the Functioning and Malfunctioning of Schools," in *Schools of Psychology: A Symposium of Papers*, ed. David L. Krantz (New York: Appleton-Century-Crofts, 1969), 1–19. See also Christopher D. Green, "Scientific Objectivity and E. B. Titchener's Experimental Psychology," *Isis* 101, no. 4 (2010): 697–721.
9. James Rowland Angell and Addison W. Moore, "Reaction-Time: A Study in Attention and Habit," *Psychological Review* 3, no. 3 (1896): 251. Angell followed up on the study, clarifying some of its central issues, in James Rowland Angell, "Habit and Attention," *Psychological Review* 5, no. 2 (1898): 179–83. For "Every one knows," see James, *Principles*, vol. I, 403. On the Chicago School, see Andrew Backe, "John Dewey and Early Chicago Functionalism," *History of Psychology* 4, no. 4 (2001): 323–40.
10. John Dewey, "The Reflex Arc Concept in Psychology," *Psychological Review* 3, no. 4 (1896): 369. For more on the paper, see Eric Bredo, "Evolution, Psychology, and John Dewey's Critique of the Reflex Arc Concept," *Elementary School Journal* 98, no. 5 (1998): 447–66.
11. Dewey, "Reflex Arc," 361.
12. For more on changing understandings of attention in this period, see Benjamin MacDonald Schmidt, "Paying Attention: Imagining and Measuring a Psychological Subject in American Culture, 1886-1960" (PhD diss., Princeton University, 2013). See also Jonathan Crary, "Spectacle, Attention, Counter-Memory," *October* 50 (1989): 97–107.
13. Leon Solomons and Gertrude Stein, "Normal Motor Automatism," *Psychological Review* 3, no. 5 (1896): 502. For more on Stein, including her work with James, see Steven Meyer, *Irresistible Dictation: Gertrude Stein and the Correlations of Writing and Science* (Stanford, CA: Stanford University Press, 2001). On Titchener's dismissal of Cattell's results, see Titchener, "Simple Reactions," 76. On Cattell, see Michael M. Sokal, "Scientific Biography, Cognitive Deficits, and Laboratory Practice: James McKeen Cattell and Early American Experimental Psychology, 1880–1904," *Isis* 101, no. 3 (2010): 531–54.
14. John Lubbock, *Ants, Bees, and Wasps: A Record of Observations on the Habits of the Social Hymenoptera* (London: K. Paul, Trench, 1882), 247.

For Thorndike's reflection, see Edward L. Thorndike, "Animal Intelligence: An Experimental Study of the Associative Processes in Animals," *Psychological Review: Monograph Supplements* 2, no. 4 (1898): 30. On his apparatus, see John C. Burnham, "Thorndike's Puzzle Boxes," *Journal of the History of the Behavioral Sciences* 8, no. 2 (1972): 159–67.

15. Thorndike, "Animal Intelligence," 6, 27, 101.
16. Edward Thorndike, "Mental Fatigue," *Journal of the American Medical Association* 34, no. 12 (1900): 727. On fatigue, see Anson Rabinbach, *The Human Motor: Energy, Fatigue, and the Origins of Modernity* (New York: Basic Books, 1990). For the table shown in figure 2.3, see Thorndike, "Animal Intelligence," 69.
17. Boring, *A History of Experimental Psychology*, ix, 268, 311. On the interplay between Fechner's theories and his lived experience, see Joshua Bauchner, "Fechner on a Walk: Everyday Investigations of the Mind-Body Relationship," *Historical Studies in the Natural Sciences* 51, no. 1 (2021): 1–47.
18. Edward Wheeler Scripture, *Thinking, Feeling, Doing* (Meadville, PA: Flood and Vincent, 1895), 56, 41. On vital principle, see E. W. Scripture, "Accurate Work in Psychology," *American Journal of Psychology* 6, no. 3 (1894): 427. On isolation, see Henning Schmidgen, "Time and Noise: The Stable Surroundings of Reaction Experiments, 1860–1890," *Studies in History and Philosophy of Science Part C: Studies in History and Philosophy of Biological and Biomedical Sciences* 34, no. 2 (2003): 237–75.
19. Scripture, "Accurate Work," 429. On enthusiasm, see Boring, *A History of Experimental Psychology*, 619.

3

PAYING ATTENTION TO THE BIRDS

Ornithologists and Listening

ALEXANDRA HUI

"These 'haunt' me."
—Clarence Stone, describing white-crowned sparrow
calls to Verdi Burtch, May 11, 1920

A SCENE OF ATTENTION: CHOPIN AND THE VIREO

F. Schuyler Mathews was a composer, watercolorist, and author of several books on trees, wild flowers, and birds. His popular *Field Book of Wild Birds and Their Music* (1904) was the last of a handful of texts written that employed musical representations of bird sounds. Mnemonics alone, Mathews explained, were "wholly inadequate, if not extremely unscientific" in representing birdsong and sufficient only to convey rhythm. Additionally, mnemonic descriptions were bound to time and place and therefore unstable.

The yellow-throated vireo, for example, was commonly described as saying, "See me? I'm here. Where are you?" During the Boer War, however, Mathews claimed that he heard the bird say, "Mafeking, Modder River, Buluwayo [*sic*], Molappo,

FIGURE 3.1 "Buluwayo" represented in music—F. Schuyler Mathews

Boer War!" With this new mnemonic, not only did the rhythm shift but the changed phonemes altered the sounds themselves. Mathews determined that one's sensory-perceptual experience of a sound of interest is informed by shifts in the background. Discussions of colonial violence in Africa altered how birds in the woods of New Hampshire sounded to Mathews.

The slurring of the yellow-throated vireo's "Buluwayo" was best represented, Mathews continued, by mnemonic-free music, as shown in figure 3.1. Notice that the bird sang in a specific key, as well as in chords. (Readers were presumed to have enough musical background to see these notes on the page and hear them their mind's ear.) If the passage looks or sounds familiar, it may be because it is similar to several bars of Chopin's *Fantaisie-Impromptu*, which Mathews also included in his *Field Book* (figure 3.2).

By placing his musical notation of bird sounds next to a well-known human composition, Mathews provided his readers with an earworm. If one of them heard the roiling sounds of Chopin while walking in the woods, it was likely to be coming from a yellow-throated vireo. Then again, the vireo's liquid calls vary.

FIGURE 3.2 Two bars of Chopin's *Fantaisie-Impromptu*—
F. Schuyler Mathews

A birder who listened only for runs of *Fantaisie-Impromptu* might miss a lot of vireos.

We should not assume that these birds were actually singing the opening bars of a Chopin composition. We should assume, however, that at least some naturalists heard these sounds. They heard pure intervals. They heard the shadings of different keys. Indeed, they couldn't hear these sounds any other way. How one thinks about sound informs what one hears. And what one hears informs how one thinks about sound. Listening attentively for something will alter what one subsequently hears.[1]

In this chapter, I widen this phenomenological looping beyond the individual. I explore how specific forms of attentive listening are conveyed to others. We can think of Mathews's *Field Book* as a multimodal training device of sorts. Mathews sought to standardize not only forms of birdsong representation but also how readers listened to birds' sounds. After studying Mathews's work, readers might have become more focused listeners, listening for specific sounds or patterns—say, a "Buluwayo" in the forest or a piano impromptu. The case study that follows is an example of how a specific form of attentive listening was, deliberately or not, conveyed to others. I examine the documentation techniques of two expert birders used while sharing birding tramps in person and via correspondence over the course of a year. By examining a possible mechanism through which individuals are trained to listen in a highly specific way, I address the more general question of how one alters, trains, or standardizes another individual's attention.

AUDITORY ATTENTION IN THE FOREST

Clarence Stone (1870–1930) and Verdi Burtch (1868–1945) were both from Yates County, New York, the towns of Branchport and Penn Yan, respectively. The two towns were, in the first decades of the twentieth century, connected by an electric trolley. Stone had a position at the Branchport Museum of Natural History but also took on seasonal work cutting trails and as a game warden. He contributed birding notes to local newspapers and books, including *The Auk*, *Bird-Lore*, and *Life Histories of North American Birds*, published by the Smithsonian Institution. Burtch operated a general store and similarly contributed to ornithological publications. He was also a skilled wildlife photographer and developed study series for the U.S. Department of Education and Pennsylvania State College and lantern slides for the U.S. Department of Agriculture. The two men began birding together in their youth and continued to do so over several decades. During the late nineteenth and early twentieth centuries, professionalization increased across the sciences, and ornithology was notable for its ongoing collaboration between professional ornithologists and avocational birders.[2] Stone and Burtch are emblematic of the latter; they were highly respected and well known for their ability to track rare birds and enthusiastic contributors to the science of ornithology through correspondence, publications, and banding efforts.

The two men would go on hours-long birding tramps together at least once a week, sometimes with other birders. Their field notebooks indicate that the group would often hike to a spot together, then split up in pursuit of specific birds and reconvene later. Stone's and Burtch's field notebooks differed significantly (more on this in a moment), but both carefully documented the location of species, the number of individuals found, behaviors, and details of nests, including their contents. At the end of the season, the men would add indexes to their field notebooks, organized by species and location, and cross-reference their notes.

Stone's field notebooks, spanning from 1887 to 1920, demonstrate that his documentation practices shifted over time.[3] In his field notebook beginning in 1905, the prose is thoughtful and the handwriting tidy and smooth, suggesting that he was writing at the end of the day upon reflection. Around 1911, however, his notes become sloppier. There are splashes of ink, inconsistent spacing, and marginalia, suggesting that he had begun to write more on the fly, possibly in the field. He also started to include maps, sketches of nest locations, a few silly cartoons, and news clippings.

Stone also consistently documented the bird sounds he heard. But his notes make it clear that he struggled to represent these sounds. He used a variety of representations, including simple descriptions (e.g., the "derisive laughter" of the sora rail), onomatopoeia (e.g., the "cac-cac-cac-cac" of goshawks), graphic representations, and musical notation. Over the decades, Stone

68 ∞ HISTORIES OF ATTENTION

FIGURE 3.3 Musical notation from Clarence Stone's field notebook illustrating two songs from a white-crowned sparrow

fiddled with different articulations of sound on the written page. The barred owl, for example, sounded like "wuh-wuh" or "wuw-wuw" or "huh-huh" in October 1920. The following year, he tried "hoo-hoo" and then settled on "a high toned hoo—hoo-hoo—hoo—hoo-hoo etc."[4]

Stone also made use of his decades of diligent field note-taking by referring back to earlier sonic encounters for comparison. On May 10, 1920, for example, he heard two songs from a white-crowned sparrow and described them in the musical notation shown in figure 3.3. Two days later, he heard a new song from this bird and described it as shown in figure 3.4.

Later, Stone returned to these pages and arranged his musical representations of the white-crowned sparrow (one from 1905 and the three from 1920) in the upper margins of two facing

FIGURE 3.4 A new song from a white-crowned sparrow—
F. Schuyler Mathews

FIGURE 3.5 All of Stone's musical notation representations of white-crowned sparrow songs

pages, noting the species in red pencil, presumably for easy reference (figure 3.5).

What would be interesting but is impossible to know from his field notebooks is whether, in 1920, he first thought back to the 1905 white-crowned sparrow in his mind's ear or referred to his index and then re-sounded the 1905 call upon reading the notation. However, we do know that he engaged with his field notebook entries at least twice, first when writing them and again when referring back to them to compare earlier with later birdsong representations. We know that Stone was working through a sensory-perceptual feedback loop, framed by the multimodal object of his field notebook. The written, graphic, and musical representations of bird sounds, the drawings, and Stone's practice of flipping back and forth through the pages of his field notebooks according to date or indexed species all contributed to an increasingly refined form of attentive listening.

I suspect that Stone moved from a broad and open form of listening (interpreting the myriad "huhs" and "woos" as coming from barred owls) toward a more focused form of listening in which he used a single song to represent an entire species ("hoo-hoo" only). His field notes show that he seemed to be actively trying to refine his hearing and settle on the best way to represent a species' sound on paper. On June 27, 1920, for example, he wrote, "Saw a Grosbeak and heard it twitter a strange note so I learned something. It has [the] quality of a Bluebird's voice and evidently has something to do with warning her young ones—an alarm of caution."[5] Because he did so much local birding, revisiting sites several times a week, he must have been aware of the individual variety of song within species. Yet Stone does not appear to have reveled in the specificity of individuals or to have delighted in the variation in a species' sounds. Although I can't be sure, I suspect he may have been trying to develop a stable auditory reference system, one that would work both in his mind's ear and on the pages of his field notebooks.

Verdi Burtch listened to the woods differently. Burtch was far more interested in vegetation and landscape than Stone, who referred to a tree or shrub only in reference to a bird or its nest. Burtch spent as much text describing with delight various bog plant species—"But O! The lure of it! The pale laurel sphagnum and pitcher plants. . . ."—as he did birds.[6] He also included rather romantic descriptions of encounters with other wildlife, such as a moment of eye contact made with a doe or the splash of a beaver paddling away. Rarely did Burtch document birds' sounds or even indicate whether they were making sounds at all.

Elsewhere I have used the term *field notebook listening* to capture the process through which an individual alters their sensory-perceptual framework: by listening attentively to a sound, contemplating it, capturing it in written form, referring back to the written representation at a later date, and re-sounding it in their auditory imagination.[7] By documenting a fleeting sound in his field notebook, Stone (and possibly Burtch) fixed it for a time in this historical record, as well as in his auditory memory and his framework for experiencing the world. That is, until a new field notebook listening experience altered his perception of the sound once more.

So, what came first, the representation of a heard sound or the attentive listening to it? It's hard to say, and certainly nothing conclusive can be drawn from the limited archival record in the case of Burtch and Stone. But my larger point is that the sensory-perceptual loop exists and that *it* is evident in the archival record. We can trace shifting representations of sounds in Stone's and Burtch's field notebooks and presume interrelated shifts in sensory experiences and the attention paid to them—contributing to the creation of altered subjective experiences of each birder's environment. Listening was a multimodal, material process for my historical actors, as it has been for this historian's reading of the historical record.

CONVEYING SOUND: NOTATION AS AN ATTENTIVE PRACTICE

Next I consider how specific forms of attentive listening might have been conveyed between Stone and Burtch. I struggled with the term *convey*. Before landing on it, I also considered *suggest*, *transfer*, and *train*—but not *impose* or *standardize*. I'd like to leave room (which already exists in the silence of the archives) for my historical actors to maintain agency over their sensory-perceptual experiences. Further, the presentation and reception of attentive practices in this case study presumably were unconscious, or at least unintentional, efforts for both parties. Still, I can trace some of the processes through which attentive listening may have been co-cultivated between the two birders.

Over the course of two weeks at the end of June 1920, Stone and Burtch took daily day-long birding tramps together, sometimes with others joining them. A comparison of their field notebook entries for these shared excursions reveals how the men listened to both their environment in general and to specific birds differently. For example, on June 20, the two identified a Nashville warbler singing in a tall birch tree set back from the road. Stone wrote, "Heard many Nashville warblers singing and their song is sharp, uttered something like the following, oft repeated," followed by a two-part squiggly line, the first portion labeled "slow" and "sharp" and the second portion labeled "rapid" and "'Chippy' like trills."[8]

Burtch's notes imply that he couldn't identify the species of warbler until he could see it. Stone, however, described the bird as a Nashville warbler from the start, noted that its song was typical of the species, and then represented the song on the page with a description and a graphic representation. Stone appears to have navigated the woods more by sound than sight and to have been capable of recalling birdsongs of the past to compare them with those he was hearing in the moment. From this comparison of notes, I think we can infer the negative of Burtch's auditory experience. We know what he didn't hear: he didn't hear a Nashville warbler. At least not until he saw it.

Later in the week of the warbler sighting, Burtch described hearing the distress cries of a pair of hawks, which reminded him of the sharp-shinned hawk's cries "but was coarser and it proved to be a pair of Goshawks."⁹ Describing the same encounter, Stone wrote that he was startled by angry goshawk cries and that "the 'cac-cac-cac-cac-cac-cac-etc.' sounded like the sharp-shins but was coarser and louder."¹⁰ Both men quickly identified the species and its relative auditory similarity to its cousin, but Stone also included a description of the birds' sound using onomatopoeia ("cac-cac-cac"). Stone's inclusion of a variety of textual representations of birdsong occurred frequently enough in his writings to suggest that the sounds were the species to him. He documented his observations by documenting the birds' sounds.

This form of identification can also be seen in Stone's letters to Burtch. On May 11, 1920, when he was wrestling with how to describe the sound of the white-crowned sparrow in his field notebook, Stone also wrote to Burtch about it. He used musical notation to describe "all I heard" on the previous day (figure 3.6), which matches what he indicated in his field notebook for May 10 (figure 3.3).

FIGURE 3.6 The song of a white-crowned sparrow as described by Stone in a letter to Burtch

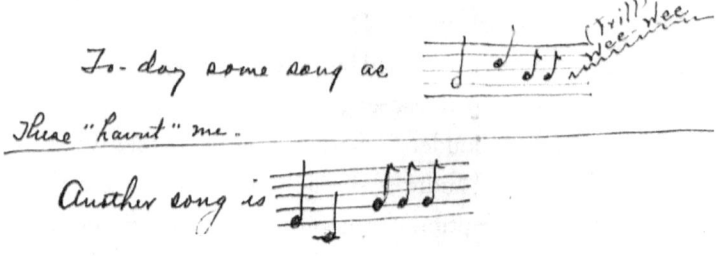

FIGURE 3.7 A second description of a white-crowned sparrow's song in a letter from Stone to Burtch

But then, on the eleventh, "some sang as" the notation illustrated in figure 3.7.

"These 'haunt' me," Stone said. Two days later, he wrote to Burtch about having spent thirty minutes the evening before listening to a hedge full of sparrows, describing their sound as shown in figure 3.8.

Here we see an expansion of Stone's field notebook listening into an epistolary medium. The reinforcement loop occurring as he wrote in his field notebook was also occurring as he

FIGURE 3.8 A new song of the white-crowned sparrow as described by Stone in a letter to Burtch

resketched the bird sound in his letters. But this time the looping also included Burtch. In his own field notebooks, Burtch noted that Stone had observed white-crowned sparrows.[11] As Burtch read Stone's letters, he would have recreated the sounds in his auditory imagination and associated them with the species.

On October 10, 1920, likely following up on an in-person conversation (there is no note of olive-sided flycatchers in either man's field notebook in the days leading up to the tenth, and the birds' southward migration had been underway since July), Stone wrote to Burtch about his notes on the olive-sided flycatcher from a previous summer.[12] He wrote that he had heard "an abundance of . . . their clear and oft repeated enunciation of 'pip-pip-pip' 'pip-pip-pip' all along the lake margins." And then, "'Pip-pip-pip' or as I heard it all this season 'puk-puk-puk' is characteristic of Olive-side[d] and no other Flycatcher." He also recounted for Burtch a field notebook entry from June 7, 1911, in which he had been "pleased to hear the oft repeated 'pip-pip-pip' of an Olive-sided Flycatcher." And then, in the current year through August, "I heard and [saw] the Olive-side[d] 'pip-pip-pip' all along the bog bordered beaver ponds." By contrast, "the yellow-bellied Flycatcher is a hermit in summer. He don't sing nor yell nor whistle like the Olive-sided . . . never 'pip-pip-pips' . . . the Yellow-bellied flycatcher voice is a low, soft, mellow, just a plaintive 'peea-peea' in Wood Pewee fashion." Apparently Stone's need to clarify to Burtch the difference between the two species' sounds was so strong that he ended with, "I'm coming down to-night to visit you at the house and talk it over."

Every time Stone mentioned the olive-sided flycatcher, he described its sound using onomatopoeia, both in his field notebooks and for Burtch in his letters. One can imagine Burtch sounding out the "pip-pip-pip" in his mind's ear each time he read another passage of it. If Stone had been trying to train

Burtch's ear to hear a specific version of the olive-sided flycatcher's sound, planting such an earworm was a good strategy. The process of attentive field notebook listening, broadly defined, might very well enable the standardization not only of how a sound is represented on paper but also how it is heard by human bodies.

Let's widen the lens a bit now. Though remarkably fastidious in their field notes and the regularity of their birding tramps, Stone and Burtch were more emblematic of the birders of their time than exceptional. In this period, the new reliability of the car; the proliferation of state and national parks, nature study, and scouting; and the promotion of feeding birds by birding organizations as a civic obligation fueled an explosion of interest in avocational bird-watching.[13] Guidebooks and new technologies such as film recordings, ear-training records, and radio programs addressing conservation advanced the science of conservation.[14] I believe that these were also technologies of attentive listening, mobilized by professional ornithologists and avocational birders alike, that were used to archive and standardize sounds—or rather, to standardize humans' hearing of sounds. I also consider the field notebook and letter to be technologies of attentive listening. Through these proliferating technologies and practices, the sounds of the forest and fen changed—at least to those listening attentively.

THE BULUWAYO, MEMORY, AND AUDITION

I originally wanted to title this chapter "Listening for the 'Buluwayo'"—admittedly a bit misleading for a paper that doesn't discuss Zimbabwe at all. But I wanted to emphasize F. Schuyler

Mathews's point that sounds specific in time can infiltrate the auditory memory, altering a person's subsequent listening experiences of those sounds. Between 1880 and 1881, Mathews heard the yellow-throated vireo saying, "Buluwayo." Then he heard it singing Chopin. By documenting the bird's song as "Buluwayo" and as a musical passage in the pages of his *Field Book*, he conveyed these unique sounds to his readers. As they moved through the woods listening attentively for them—by the very act of listening for "Buluwayo" or *Fantaisie-Impromptu*—they heard them.

In this chapter, I have examined the birding practices of two avocational birders in the early decades of the twentieth century. Their careful documentation of their birding tramps shows shifting, then stabilizing forms of birdsong representations, particularly for Stone. We can presume that these shifts in representation also indicate shifts in what Stone heard—and what Stone listened for. That is, we can see in the pages of his field notebooks how Stone's attentive listening altered his soundscape. We can also see how the sounds heard by Stone might have begun to resound in Burtch's ears as well. Stone's efforts to convey his way of listening to Burtch were likely not deliberate but rather a consequence of the nature and medium of their interactions—tramps together and then diligent documentation of their own and the other's observations.[15] The observation of the Nashville warbler indicates that Stone saw with his ears, whereas Burtch heard with his eyes. Perhaps because the two birders occupied two extremes of an auditory-visual and sensory-perceptual framework, paper was an especially effective medium for conveyance. I would hazard that because Burtch was not a particularly attentive listener, Stone was compelled to write out the sounds he heard for Burtch in his letters. The cultivation of attentive listening in others was a process aided by writing and drawing in field notebooks and letters: *field notebook listening*. As this process of representation (and re-representation) unfolded over time, specific sounds were reinforced and standardized in Stone's and Burtch's auditory memories. As the sounds archived in the men's mind's ears shifted over time, so too did their subsequent auditory experiences. The sounds altered their soundscapes. Attention is world-making.

Though it would be difficult to establish from the archival record, we can consider whether attentive listening reflected and reinforced on paper drives the standardization of sounds both for and between individual listeners. Or, put slightly differently, does the use of images and words to represent auditory experiences codevelop with other idealist, universalizing goals of modern science such as type specimens? If so, a close examination of the sensory-perceptual curve of the feedback loop between attentive listening and the representation of it could offer paths not yet taken.

Considering what does *not* get documented in a field notebook decenters the human actors from this story and instead hints at a multispecies history. If we invert the standardization story, if we turn away from it and direct our attention to the background, what thrumming variation might we find? Surely a cacophony of voices sounded in the woods, but, as far as we can know, it went unheard.

NOTES

1. This phenomenon in hearing, termed accommodation, has been studied by psychoacousticians since the nineteenth century. Want to experience it yourself? Listen to a piece of music. Now, listen to it again, but focus on a particular instrument, perhaps the one that carries the melody the most. Did it sound different that time? What if you listen to the piece again while listening attentively to just the bass line or

just the percussion? Congratulations! By altering your sensory-perceptual experience through directed attention, you have experienced accommodation.

2. Mark V. Barrow Jr., *A Passion for Birds: American Ornithology After Audubon* (Princeton, NJ: Princeton University Press, 1998); Thomas R. Dunlap, *In the Field, Among the Feathered: A History of Birders and Their Guides* (Oxford: Oxford University Press, 2011); Nancy J. Jacobs, *Birders of Africa: History of a Network* (New Haven, CT: Yale University Press, 2016); Daniel Lewis, *The Feathery Tribe: Robert Ridgway and the Modern Study of Birds* (New Haven, CT: Yale University Press, 2012); Stephen Moss, *A Bird in the Bush: A Social History of Birdwatching* (London: Aurum, 2004).

3. I have looked only at the notebooks from 1905 to 1920, which are located at the Cornell University Archives. I was unable to access earlier notebooks (1885–1903) at the Cornell University Archives and the 1887–1896 notebooks at the New York State Library at the time of writing because of travel restrictions related to the COVID-19 pandemic.

4. Clarence Stone, field notebooks, April 27, 1921, 184, box 3, folder 15, Verdi Burtch Papers, Cornell University Archives.

5. Clarence Stone, field notebooks, June 26, 1920, 141, box 3, folder 15, Verdi Burtch Papers, Cornell University Archives.

6. Verdi Burtch, field notebooks, June 20, 1920, 175, box 3, folder 16, Verdi Burtch Papers, Cornell University Archives.

7. Alexandra Hui, "From 'Wuh Wuh' to 'Hoo-Hoo' and the Rituals of Representing Bird Song, 1885–1925," in *Objects and Standards: On the Limitations and Effects of Fixing and Measuring Life*, ed. Tord Larsen et al. (Durham, NC: Carolina Academic Press, 2021), 231–54.

8. Clarence Stone, field notebooks, June 20, 1920, 135, box 3, folder 15, Verdi Burtch Papers, Cornell University Archives.

9. Verdi Burtch, field notebooks, June 28, 1920, 192, box 3, folder 16, Verdi Burtch Papers, Cornell University Archives.

10. Clarence Stone, field notebooks, June 28, 1920, 141, box 3, folder 15, Verdi Burtch Papers, Cornell University Archives.

11. Verdi Burtch, field notebooks, May 13, 1920, 137, box 3, folder 16, Verdi Burtch Papers, Cornell University Archives.

12. Clarence Stone to Verdi Burtch, October 10, 1920, 152–53, box 3, folder 15, Verdi Burtch Papers, Cornell University Archives.

13. Barrow, *A Passion for Birds*; Etienne Benson, "A Centrifuge of Calculation: Managing Data and Enthusiasm in Early Twentieth-Century Bird Banding," *Osiris* 32, no. 1 (2017): 286–306; Dunlap, *In the Field*.
14. Joeri Bruyninckx, *Listening in the Field: Recording and the Science of Birdsong* (Cambridge, MA: MIT Press, 2018); Alexandra Hui, "Listening to Extinction: Early Conservation Radio Sounds and the Silences of Species," *American Historical Review* 126, no. 4 (2021): 1371–95; Rachel Mundy, *Animal Musicalities: Birds, Beasts, and Evolutionary Listening* (Middletown, CT: Wesleyan University Press, 2018).
15. According to modern bird-watching conventions, from data collection to competitive birding counts, hearing a bird counts as observing it. Notice how the language nevertheless prioritizes the visual.

4

ATTENTION, ART, AND PSYCHOTHERAPEUTICS

JULIAN CHEHIRIAN

THE DREAM

In 1890, in the Russian Empire, or in the placeless matrices of the nocturnal mind, four-year-old Sergei dreams a dream. He is awake and in his bed—in a bed beneath a window on a wintry night. Beyond the window lies an old row of walnut trees. Abruptly, the window swings open, and the distance between Sergei and the trees thins. A row of six or seven white wolves appears across the branches of a walnut tree. Their ears are pricked up and they stare into him.

Outside of this dream space the nurse of the young aristocrat, hearing his terror-stricken sounds, rushes to awaken him. Twenty years later in Vienna, Sergei, the "Wolf Man," recounts this dream to Sigmund Freud (1856–1939) in Freud's consulting room. And Freud, who was otherwise known to dismiss his patients' attempts to articulate their dreams in visual terms by drawing or painting, allows this patient to do so. Sergei Pankejeff drew his scene of somnium (figure 4.1).

Freud, who is known for developing the "talking cure," believed that the unconscious traffics in images. "We experience it [a dream] predominantly in visual images," he wrote. "Part of

82 ᴂ HISTORIES OF ATTENTION

FIGURE 4.1 "How did the wolves get up in the tree?" Sergei Pankejeff, 1916, pencil on paper.

the difficulty of giving an account of dreams is due to our having to translate these images into words. 'I could draw it,' a dreamer often says to us, 'but I don't know how to say it.'"[1]

Freud, however, did not solicit drawings during his sessions. Instead, he had patients speak of their dreams. This repression of the visual was part of his commitment to a very particular way of paying attention. In the 123 years since the publication of *The Interpretation of Dreams*, we've managed to overlook something fundamental to the clinical practice of psychoanalysis. It is not really about "talking", but about "attending." It is really about the psychotherapeutic application of attention. Attention is the means by which psychoanalysis is practiced. Attention is the method of psychoanalysis, but it is also the object of psychoanalysis.

ATTENTION, ART, AND PSYCHOTHERAPEUTICS ⌘ 83

If we read across the broader history of the psychotherapeutic sciences of mind and behavior with an eye to prescriptions for attention, we will find that practices ranging from moral therapy to hypnosis to art therapy have attempted to get at the mind *by way of* attention, which practitioners have sought to redirect, reorganize, or transform. In order to locate attention's operative role between practitioners and clinical subjects, we must examine practices and protocols that guide practitioners and patients towards particular ways of attending.

This chapter considers how Freud understood attention and how he made use of it in his clinical practice. Beginning and ending with images produced in historically divergent therapeutic contexts, it traces a post-Freudian shift in psychotherapeutic practices of attention away from analysis and interpretation through language, and towards efforts to bypass language and apprehend the non-verbal. Traveling along a seam between word and image, it attempts to locate the place of attention in the emergence of art therapy, considering how practices of attending after Freud rendered new modes of subjectivity, representation, and interpretation.

ATTENTION

The idea of "attention" and attentiveness emerged in the nineteenth century as a way to make sense of a human subject who was no longer considered to be autonomous: one not governed by reason for whom the world was no longer transparently available to the senses.[2] Attention cropped up in various spaces. One of these sites, as Henry Cowles described in chapter 2, was the psychology laboratory. But attention was an object of inquiry and concern outside the laboratory, too. Social critics like Reverend

J. H. McIlvaine saw the mass press as entrapping voluntary attention by appealing to base instincts. Antebellum education reformers were concerned with the "psychic damage wrought by modernity" and imagined the ideal setting for a school reformatory to be the countryside—far from the increasing industrialization and social decay of the city. Caleb Smith writes that in this "environmentalist diagnosis of the psychic symptoms of modernity, and in the fantasy of rehabilitation that accompanies it, there is a distinctly modern, secular theory of distraction and attention."[3] Distraction as pathology.

Attention also found itself at the edge of the science of psychology: in spaces devoted to the treatment of the mentally ill. Asylums, unlike their predecessors (madhouses), were designed with therapeutics in mind. But their method was essentially organizational. Patients were believed to be susceptible to pathologies of normal, rational, sane thought. Asylums aimed to *organize*, redirect, and give shape to the attentive faculties of the mentally ill. Their superintendents, who had no medical credentials, were chiefly preoccupied with padding patients' schedules and creating workshops to occupy their hands and minds—or, as I argue, their attention. Asylum architecture and management bespoke a central aim: to appeal to a rationality inherent to the individual by way of order, encouraging "self-discipline, limit setting and focusing."[4] In short, asylum superintendents prescribed models for the self-management of attention.

The idea that one's attention is used to manage broad strata of experiences was foundational to early scientific discourses on attention. William James (1842–1910), a professor of philosophy at Harvard, wrote the most influential account of attention of his time. James theorized that "volition is nothing but attention."[5] Attention for him was selective, focal, and exhaustible. Beyond attention is distraction: a state of vacuity or negative reverie, a momentary loss of volition but one pregnant with its own end. According to James, what brings about that end is unclear: "An energy is given, something—we know not what." When one is distracted, "attention is dispersed." One is taken by a "sort of solemn sense of surrender to the empty passing of time." "We float with it," with this "curious state of inhibition," with this "inattentive dispersion of mind." Distraction is not useful. An effort is required to overcome the solicitation of our attention by objects of interest. According to James, *to sustain attention is to overcome the resistance of the wayward flow of associations that lead us away from objects of our volition.*

Psychoanalysis was the first form of psychotherapy built explicitly around attention. But it may be the only method in the mind sciences founded on a tactical employment of distraction. Freud is not a name that readers are likely to associate with theories of attention. But he belongs in the conversation. He developed a theory of attention that is eminently non-Jamesian: distraction is not the opposite of attention, and it is no less than the means of accessing the truth of the mind. For particular reasons, it is the only way to do so. The major insight of psychoanalysis is that the mind is not available to itself to freely attend to. Attention is impeded by psychic repression, censorship, and the unconscious mind. But attention is not just part of the theory of mind used in psychoanalysis. It is wielded in the very practice of psychoanalysis. Attention is the *method*, but it is also the *object*. Pause on that. Attention to people is the method of the talking cure, but it is also its object: the attention of the patient. Psychoanalysis took place in a sanctuary within the city; it consisted in a specific relationship between two individuals in a new sort of space: the private cabinet of a neurologist-turned-psychoanalyst.

In his early and long-unpublished *Project for a Scientific Psychology* (1895), Freud gave an account of the "mechanism of physical attention." He insisted that "attention is biologically justified," serving to mediate sensory input and delimit objects of interest.[6] In his earlier, mechanistic model of the psyche, he considered this mediation to operate on the basis of his drive theory, for which libidinal energy was believed to be a significant force arising in the form of "cathexis"—an investment or attachment to psychological objects. Freud denoted attention as a psychical "function," one that is *exercised*, is an element of "becoming conscious," and can be "diverted," suggesting its finite and focal qualities. At this stage, he sounded Jamesian.

But as he developed his thinking on the role of the unconscious and its bearing on our models of the psyche, he diverged from James. Attention was no longer coextensive with volition. It had just as much to do with structural nonvolition. One attends to both less and more than what one is aware of attending to. Attention is not a transparent instrument of our will but rather is mediated. The "selective" and "focal" qualities of attention are complicated by psychological mechanisms that govern what one attends to and to what degree. The objects of one's attention are subject to latencies, repressions, and transferences. Freud's view of attention introduced restrictions—a complexity to volition and a catacomb to attention itself. What is accessible to and by one's attention is influenced by processes beyond the reach of one's conscious attention. If attention for James connoted a selective activation of conscious awareness, the psychoanalytic concept of attention introduced a backstage to it. Beyond these universal structures, attention is shaped by its own history—by the attentional history of the individual.

The work of psychoanalysis is to mobilize, through a practice of attention, boundary work among the conscious, the preconscious, and the unconscious. It is to identify and unearth conflicts, the effects of which upon the conscious mind are pathological symptoms. Surfacing the history of a subject's attentional formation is the *work* of psychoanalysis—to excavate the past in order to inspect its canalization of present attention and, ideally, to transform the structure. James was not concerned with the history of one's attention. He argued that something inaccessible to attention was not significant enough to capture attention: "We cannot deny that an object once attended to will remain in the memory, whilst one inattentively allowed to pass will leave no traces behind."[7] For Freud, however, attention both to the present and to memory was troubled by the stratification of consciousness.

Freud's diminishing interest in accounting for the physical mechanisms of attention coincided with his growing interest in the *application* of attention toward therapeutic ends.[8] The following passage from *The Interpretation of Dreams* can be regarded as a bridge from his mechanistic project to his attentional practice in the mobilization of psychoanalysis as a therapeutic modality: "The course of our conscious reflections shows us that we follow a particular path in our application of attention. If, as we follow this path, we come upon an idea which will not bear criticism, we break off: we drop the cathexis of attention. Now it seems that the train of thought which has thus been initiated and dropped can continue to spin itself out without attention being turned to it again, unless at some point or other it reaches a specially high degree of intensity which forces attention to it."[9]

William James, who was skeptical about the quantitative streak in the "new psychology," was intrigued by psychoanalysis. After meeting with Freud at Clark University in 1909, however, his enthusiasm was blunted. "I strongly suspect Freud, with his dream theory, of being a regular hallucine. But I hope that he and his disciples will push it to its limits," he wrote.[10] We are trying.

If psychologists discovered attention, and if social reformers and asylum planners aimed to reform and reorganize it, Freud and other psychoanalysts would go further: honing their attention to *intervene upon and transform* the attention of their therapeutic subjects. To understand this practice, we must traverse that great ravine of attention, distraction, and that great ravine of modernity, fin-de-siècle Vienna.

GLEICHSCHWEBEND

The first half of the twentieth century saw the precipitous rise of the idea that mental illness is rooted in *meaning* and that recovering meaning and making it intelligible could bring healing. This idea was most influentially developed by the field of psychoanalysis. Much psychoanalytic work involves the investigation of meaning that has been repressed or displaced by unconscious processes. This requires a specific *form* of attention, referred to by Freud as *gleichschwebend*, typically translated as "free-floating" or "evenly distributed attention." But to understand it, we must understand another key piece of the puzzle: free association.

In 1900, the same year that *The Interpretation of Dreams* was published, Arthur Schnitzler (1862–1931) described in his novel *Lieutenant Gustl* a method of free association used to follow the attentional path of an Austrian officer on the eve of a duel:

> How much longer is this thing going to last? Let's see what time it is . . . perhaps I shouldn't look at my watch at a serious concert like this. But no one will see me. If anyone does, I'll know he's paying just as little attention as I am. In that case I certainly won't be embarrassed. . . . Only quarter to ten? . . . I feel as though I'd been here for hours. I'm just not used to going to concerts. . . .

What's that they're playing? I'll have a look at the program. . . .
Yes that's what it is: an oratorio. Thought it was a mass. That sort
of thing belongs in church. Besides, the advantage that church has
is that you can leave whenever you want to. I wish I were sitting
on the aisle![11]

As Lieutenant Gustl fingers his coat pockets in the darkened opera house and unfolds his printed program, some convergence of associations surfaces other moments in his attention, perhaps of entrapment or malaise, and soon after of things distant from those thoughts. His freely wandering attention, uncaptured by the oratorio, slides about in the wake of internal and external impressions. The lieutenant's attention flits and skates, moves laterally, and accepts solicitations from his memories. Mario Erdheim, writing on the genealogy of practices of free association in the arts and sciences, observes that such a "state of 'evenly floating attention' . . . usually occurs in relation to objects" but is "extremely difficult to maintain as soon as subjects appear."[12] A diffuse, noncommittal style of attending to subjects would "run counter to everything that is taught to us in an average upbringing, namely to align ourselves in a positive or negative sense with other people and to be specifically attentive." Freud was affected by the novel and confided in a letter to Schnitzler on the occasion of Schnitzler's sixtieth birthday in 1922 that "he had avoided a meeting out of fear of facing his double [*Doppelgängerscheu*]."[13]

Freud's most significant contribution, Mario Erdheim argues, was not the discovery of an unconscious—already a "central theme of Romanticism and then of Viennese Decadence." Rather, it was his development of "a specific and completely new attitude towards the unconscious."[14] Freud developed a method of *sustaining free association in relation to subjects*. This he achieved as part of an attentional protocol in the setting of individual psychoanalysis, in which the patient "is supposed to say more than what his censorship allows (bracing 'logic, morality, and conscious objectives'), while the analyst is supposed to hear more than what his theory, his previous practice and his intentions allow him."[15] Freud's method of free-floating attention activated an approach of listening, with and through one's unconscious, to the unconscious of another. More profoundly, his major achievement, Erdheim suggests, was the activation of techniques of free association—of an evenly distributed attention—within a *social relationship*. What was unique about this process was that "in and through it the unconscious was supposed to be made conscious." The relationship between analyst and patient was intended above all to be a therapeutic one.

Freud recommended this method of attention toward patients for two reasons. First, "it saves tension, which, after all, it is not possible to maintain for hours." Second, "it avoids the dangers that threaten in the case of deliberate attention directed towards a particular aim."[16] In the case of his work with the father of a young boy known as "Little Hans," Freud called attention to the *deferred temporality* of insight in analysis: "There is no sense in which our task is to 'understand' a medical case straight away; this can only happen later, once we have formed sufficiently clear impressions of the subject." One must "suspend judgement for the time being and consider all the relevant material with the same degree of attention."[17]

If an analyst acts in accordance with his conscious expectations for the session or the predicament of the patient, "he is in danger of never finding anything but what he already knows; and if he follows his inclinations he will certainly falsify what he may perceive."[18] Freud advised that "the attitude which the analytic physician could most advantageously adopt was to surrender himself to his own unconscious mental activity, in a state of evenly suspended attention, to avoid so far as possible reflection and the construction of conscious expectations . . . to catch the drift of the patient's unconscious with his own unconscious."[19] Involuntary attention, willed.

Theodor Reik (1888–1969), one of Freud's first students, examined Freud's ideas about free-floating attention in his book *Listening with the Third Ear* (1948). Reik repeated Freud's concern: if we do not attend freely, "we may never find anything but what we are prepared to find."[20] "Attention, we have always been taught, implies selection," he wrote. But there is also danger in selection. Insofar as it enables immediate reaction, selective attention "cannot be the kind of attention used in analysis because we often do not recognize the psychological significance of the things that we are told until afterward." As such, the analyst's attention should be "directed to preparation for subsequent understanding." Free-floating attention builds up "a storeroom of impressions, from which later knowledge will suddenly emerge."[21] This statement recalls the historian Carlo Ginzburg, who, in his essay on interpretation and conjectural knowledge, refers to the ancient and technical (venatic) knowledge of sniffing, interpreting, and classifying tracks or traces of what one is hunting as a "rich storehouse of knowledge."[22] The temporality of free-floating attention, then, is one of deferral. Thus, attention that wanders seemingly has an important and instrumental function: to inhibit one's tendency to reduce one's field of attention to that which appears interesting or significant. Was Freud's inhibition successful? I ask that you follow me, in the style of free-floating attention, on a bit of a diversion—a distraction.

DISTRACTION

> "The only piece of action in the dream was the opening of the window; for the wolves sat quite still and without any movement on the branches of the tree, to the right and left of the trunk, and looked at me. It seemed as though they had riveted their whole attention upon me.—I think this was my first anxiety dream."
>
> —Sergei Pankejeff

Freud came to believe that Sergei's dream was a "screen"—a distortion of a disturbing early childhood experience: witnessing a "primal scene" between his parents. The window "opening" was a distortion of Sergei's awakening, the tree a symbol of voyeurism.

But Carlo Ginzburg rereads Freud's analysis from a different angle.[23] While studying Slavic folklore, he came across several shamanic sects whose followers identified as werewolves. They claimed to have been born "with the caul" (with the amniotic membrane enclosing the fetus) or "during the twelve days between Christmas and Epiphany." From Freud's account, we know that Sergei Pankejeff was "a Russian, that he had been born with the caul, that he had been born on Christmas day." His nurse was described by the patient as a "pious and superstitious woman" to whom he was deeply attached and who read him stories.

Ginzburg believes that he has found a hermeneutic channel that Freud missed. Freud was in "full command" of his everyday cultural context and "capable of deciphering literary and other allusions, some of a hidden, innermost nature."[24] But he did not recognize an element of foreign folklore present in the dream. Specifically, he did not pick up on similarities with the fable of "The Imbecile Wolf" in the collection of the Russian Slavist Alexander Afanasyev. Instead, Freud had plumbed the (understandably) culturally proximate (German) Brothers Grimm tale, "The Seven Little Kids" (of whom six were eaten).

Pankejeff asked Freud why there were six or seven wolves in his dream. But in his drawing, he answered his own question. His unconscious, which—following Freud—operated in images, drew *five* wolves. Freud did not make much of this discrepancy between image and speech. According to Freud's theory that the wolves represented Sergei's mother and father having sex, the fact that there were more than two was the mind's way of distancing Sergei from the true nature of his terror.

Ginzburg's Diagnosis

"The cultural context behind the dream was ignored: what remained was only the individual experience, reconstructed through the network of associations deduced by the analyst."

Ginzburg's Conclusion

"In the wolf-man's nightmare we discern a dream of an initiatory character, induced by the surrounding cultural setting, or, more precisely, by a part of it. Subjected to opposing cultural pressures (the nurse, his English governess, his parents and teachers), the wolf-man's fate differed from what it might have been two or three centuries earlier. Instead of turning into a werewolf, he became a neurotic on the brink of psychosis."

Was Freud unable to escape the associative enclosure of his own cultural setting, as Ginzburg suggests? Where might the trail of interpretation have led Freud, had he worked with the dream drawing itself? How might his clinical assessment of Pankejeff have differed had he considered the drawing an *inscription* of the unconscious, rather than a refraction of it?

During the years when Freud was carrying out his work with the Wolf Man (1910–1914), an unusual Freudian was in the making. Across the Atlantic, Margaret Naumburg (1890–1983) was cultivating a different form of free-floating attention at the convergence of art, experimental pedagogy, and psychoanalytic ideas. In 1912, Naumburg was awarded a bachelor of arts from Barnard College, having studied under the educator John Dewey. That same year, she read one of the first papers about Freud published in the United States. She "did not realize, as yet, how deeply this psychoanalytic approach to the unconscious had won a response in [her] own unconscious."[25]

ATTENTION WITH IMAGES

In 1941, in the Washington Heights neighborhood of Manhattan, "John," a ten-year-old child, was admitted to a room. The room resembled an ordinary elementary school art classroom, its perimeter lined with cabinets and containing cubby shelving and stations bearing clay, paper, and watercolor and acrylic paints. John, however, was not at school. Unable to pass the first grade, recurrently truant, and reportedly unable to concentrate or "follow the teacher's instructions," he had been ejected by school officials. He was considered "almost impossible to teach as he lacked concentration, was inattentive and showed a tendency to annoy other children." John was at the New York State Psychiatric Institute and Hospital and one of a series of patients under observation as part of Margaret Naumburg's special research project investigating "the possible use of creative art as an aid in diagnosis and therapy."[26]

The room John entered was not a typical art classroom in a further sense. Naumburg would not instruct him on artistic methods or provide him with models, themes, or items around which to organize his efforts. Rather, over the course of six months, she encouraged John to produce images from his imagination in what became a jointly beheld procession of representations guiding their interlocution. John produced a series of drawings, paintings, and clay models, which Naumburg regarded as the propellant of their endeavor: to establish a space of shared attention with a child whose capacity to attend was seemingly missing in action. John's attention emerged as a central issue, yet his attention also emerged as the primary mechanism for his reformation. For it was through a practice of "art therapy," observable here in its historical nascency as method, that a project for joint attention was pursued. As John worked, he sometimes spoke about his creations. Naumburg listened and sometimes asked questions. She wrote that John's choice of materials and the formal and emotive qualities of his work could bypass and transcend the expressive limitations of his verbal communication. In her clinical view, his creations were inscriptions of his psychological state.

Naumburg took Freud at his word that the unconscious operates through images. She also believed that the unconscious could be reached by the right form of attention on the part of both analyst and patient. But, unlike Freud, she developed an approach to psychotherapy based on the idea that her patients could more effectively bypass psychic censorship and convey their unconscious experience through "spontaneous visual expression." The premise of Naumburg's art therapy was that spontaneous visual expressions conveyed contents of the unconscious mind more directly and transparently than words could. Artistic expression provided a more direct channel of communication. Without the translation and conscious or subconscious obfuscation of language. Of one case study, Naumburg wrote, "Before Mrs. Arnstein had been able to work through the 'free associations' to this complex picture, she had seen no therapeutic value in the process of art therapy as a means of uncovering her repressed conflicts. Now, however, for the first time the patient said that she was convinced that her pictures could say things before she could put them into words."[27]

Margaret Naumburg activated Freud's therapeutic attention with a key modification: introducing objects as intermediaries between analyst and patient. She aimed to circumvent psychic defenses and distortions by having patients draw their unconscious content. During the 1940s, Naumburg's intervention coalesced as the bedrock of a discipline that would gain a national association by 1969 and acknowledge her as a founding figure.

FLOATING

In 2013, I joined an art class at a psychiatric hospital as an assistant. I got to know an older patient over the course of several months. He was sixty; I was twenty-two. For the most part, he was quiet, except for occasional outbursts on apocalyptic themes. His favorites were the Third Reich, the ongoing Syrian civil war, and several conspiracies against him. "Sergei" (a pseudonym) was born in Ukraine, some 474 kilometers from, and sixty years after, the place Sergei Pankejeff had been born.

There were four other chairs in the room. But I sat next to Sergei. He drew, and I drew. Sergei's work was agitated and ripe with conflict. In his signature style, he pressed a ballpoint pen to a sheet of paper and would not lift it until his thought-image was complete. His dense, knotted lines rendered Klansmen burning crosses, Nazis shattering storefronts, and troops advancing on Moscow.

One day I asked him a question, but he seemed to ignore me. On a piece of paper, I wrote something for him in Bulgarian, which made just enough sense for him to understand. Sergei's eyes widened. For the first time, he turned to look at me with focused attention.

In the coming weeks, he drew, and I drew. And we began to talk throughout the sessions. Sergei began to draw differently. His agitated themes withdrew like a tide. The turning point seemed to be when I asked him about where he had grown up, about his hometown. He was not impressed by the question, but then I asked him if he could draw a map.

Sergei began to draw an intricate one. As he arranged the spaces, with their buildings and courtyards, I followed his pen with my eyes, much as the camera follows a path across maps in the film *A Walk Through H* by Peter Greenaway. The map was, as the title denoted, a "Schematic Portion of Rusanovsky Massif Living Quarters for Ukrainian Civilians." At the center of the schematic was a complex of nine-story apartment buildings, and in it, the apartment of his childhood. At the center of the complex was a playground. Directly to the south was the Krakov movie theater. To their north: sixteen-story apartment buildings, rendered smaller than those in Sergei's complex, perhaps because of their reduced significance to him. At the southernmost point was the Gastronom, "a government food store." Directly west of Sergei's complex was an "embankment road for vehicles" and beyond it a "riverfront park and beach" on the banks of the Dnieper.

Sergei's outbursts on the ward decreased. He was still very ill, but his attention had shifted. I think it was because attention had been extended and received. Jointed, reciprocated.

Sergei showed me a new drawing the week after he had drawn his map. It was of an aristocratic nineteenth-century couple in

FIGURE 4.2 *The Country Estate (Sergei's gift to the author)*

an elaborate horse-drawn carriage advancing down a tree-lined road in the Ukrainian countryside. I think Sergei Pankejeff's family may have looked something like this before they left for Vienna. It was peaceful. It was titled "Sunday Morning."

On my last day at the hospital, Sergei handed me a drawing and said, "I want you to keep this one."

NOTES

1. Sigmund Freud, *New Introductory Lectures on Psychoanalysis*, vol. 15, trans. and ed. James Strachey (London: Hogarth, 1963), 90.
2. Jonathan Crary, *Suspensions of Perception: Attention, Spectacle, and Modern Culture* (Cambridge, MA: MIT Press, 2001). See also Jonathan Crary, *Techniques of the Observer: On Vision and Modernity in the Nineteenth Century* (Cambridge, MA: MIT Press, 1990).
3. Caleb Smith, "Disciplines of Attention in a Secular Age" *Critical Inquiry* 45, no. 4 (2019): 884–909.
4. Edward Shorter, *A History of Psychiatry: From the Era of the Asylum to the Age of Prozac* (Hoboken, NJ: Wiley, 1998), 33. See also Carla Yanni, *The Architecture of Madness: Insane Asylums in the United States* (Minneapolis: University of Minnesota Press, 2007).
5. William James, "Attention," in *The Principles of Psychology* (1981).
6. Sigmund Freud, "Project for a Scientific Psychology," in *The Complete Psychological Works of Sigmund Freud*, vol. 1, *Pre-Psycho-Analytic Publications and Unpublished Drafts (1886–1899)* (1950), 360.
7. James, "Attention," 403–404.
8. Katja Guenther observes that, in his later writing, Freud departed from explanations of psychic trauma as connected to physical brain lesions. See Katja Guenther, *Localization and Its Discontents: A Genealogy of Psychoanalysis and the Neuro Disciplines* (Chicago: University of Chicago Press, 2015), 93.
9. Sigmund Freud, *The Interpretation of Dreams*, trans. and ed. James Strachey (1900; repr., 1955), 590.
10. Robert I. Simon, "Great Paths Cross: Freud and James at Clark University, 1909," *American Journal of Psychiatry* 124, no. 6 (1967): 831–34.

11. Arthur Schnitzler, "Lieutenant Gustl," in *Plays and Stories: Arthur Schnitzler*, trans. Richard L. Simon, ed. Egon Schwarz (London: A&C Black, 1982).
12. Mario Erdheim, "On the Problem of Free-Floating Attention," *Psyche* 42, no. 3 (1988): 222.
13. Sigmund Freud, *The Letters of Sigmund Freud*, trans. Tania Stern and James Stern, ed. Ernst L. Freud (New York: Basic Books, 1960), 339–40.
14. Erdheim, "On the Problem," 221.
15. Erdheim, "On the Problem," 223.
16. Theodor Reik, *Listening with the Third Ear: The Inner Experience of a Psychoanalyst* (New York: Farrar, Straus, 1949), 157.
17. Sigmund Freud, *The "Wolfman" and Other Cases*, trans. Louise Adey Huish (London: Penguin, 2003), 55.
18. Sigmund Freud, *Recommendations to Physicians Practising Psycho-analysis* (London: Hogarth, 1912), 112.
19. Sigmund Freud, *Two Encyclopaedia Articles* (London: Hogarth, 1955), 239.
20. Reik, *Listening*, 157.
21. Reik, *Listening*, 161.
22. Carlo Ginzburg, "Clues: Roots of an Evidential Paradigm," in *Clues, Myths, and the Historical Method* (Baltimore, MD: Johns Hopkins University Press, 2013).
23. Carlo Ginzburg, "Freud, the Wolf-Man, and the Werewolves," in *Clues, Myths, and the Historical Method* (Baltimore, MD: Johns Hopkins University Press, 2013), 132–40.
24. Ginzburg, "Freud," 134.
25. Amey Hutchins, "Margaret Naumburg Papers Finding Aid," 2000, Kislak Center for Special Collections, Rare Books and Manuscripts, University of Pennsylvania.
26. Margaret Naumburg, "A Study of the Art Work of a Behavior-Problem Boy as it Relates to Ego Development and Sexual Enlightenment," *Psychiatric Quarterly* 20, no. 1 (1945): 74.
27. Margaret Naumburg, *Dynamically Oriented Art Therapy: Its Principles and Practices—Illustrated with Three Case Studies* (Chicago: Magnolia Street, 1987), 129.

II

PHILOSOPHIES OF ATTENTION

5

ATTENTION

Mechanism and Virtue

CARLOS MONTEMAYOR

The collective theater we share is staged by what we decide to pay attention to. Our distractions are as powerful as our interests.

THE MATERIALITY OF ATTENTION

Attention stages a theater of goals, preferences, and values. In this theater, as N. F. Dobrynin said, "it is possible to look and not see . . . to listen and not hear."[1] Attention is a limited resource, so its distribution and orientation have profound social effects. *Une loge*, the painting by Louis-Léopold Boilly shown in figure 5.1, captures the possibilities and difficulties of controlling, administering, and paying attention. Some characters are paying focused attention to one another, others are bored, and still others are fighting with one another. Most seem transfixed, but very few are paying attention to what is happening on the stage.

Attention sets the stage for the theater of our lives and ultimately determines who we are. A healthy life is built upon good attention practices or routines; a harmful one is driven by detrimental attentive interests. Too much or too little attention makes a big difference. The language of *virtue*, employed in what

FIGURE 5.1 *Une loge, un jour de spectacle gratuity.*
Louis-Léopold Boilly, 1830, Musée Lambinet, Versailles, (The Picture Art Collection/ Alamy Stock Photo).

follows, is meant to capture the positive effect that attention has on our lives. Virtues make our lives good. There is no moralistic or universal ethical assumption here. The idea is simple: how you control and orient your attention has a deep impact on your life. If it helps you meet your goals, it has a beneficial effect; if it blinds you with distractions from something important, it is detrimental. I argue that attention can perform these roles automatically and through habituation—as a mechanism and as a kind of virtue or vice that requires personal involvement.

When we pay attention, we concentrate our minds on what is salient to us. We stretch our minds outwardly or introspect inwardly. We can pay attention together, and, in many cases of successful perception and communication, we must pay joint attention. We focus on a color or smell and automatically inhibit or filter out other aspects of the environment. Sometimes, attention maximizes our control over a situation, though occasionally we may become completely absorbed. Attention comes in many forms, and all are essential for intelligent behavior. We use attention to pay taxes, enjoy music, appreciate art, and understand the complex feelings of a family reunion.

What is attention, and how does it make us intelligent? In this chapter, I offer an account of attention based on my recent research, which draws on findings in philosophy and psychology.[2] I also offer a critique of our contemporary approach to attention in industry and academia. More saliently, I am concerned with how attention is employed in our culture and society more broadly: in the information age, we have lost our appreciation for what I call the *autonomy of attention*, or the need to direct, select, and maintain control over the objects of our interest. To deepen this critique and to elucidate my position, I also offer some highlights from the history of the philosophy and science of attention.

DEFINING ATTENTION AND INTELLIGENCE

Attention allows species to satisfy their needs intelligently. In this sense, it is a kind of virtue that fosters knowledge and understanding.[3] Attention has descriptive and normative dimensions. It can be described in terms of the information or computations that it processes, the areas of the brain responsible for distinct types of attention, and the behavior it produces. In terms of normative value, some kinds of attention routines are *good* or better than others, and agents *should* use these types of attention to satisfy their needs. Attention requires a certain degree of autonomy to be fully beneficial. We could not develop a healthy perspective on the world with habits of attention that are mere mechanical repetitions or knee-jerk reactions. However, the Scientific Revolution promoted a mechanical understanding of the mind. The view of attention as a manifestation of a person's character fit oddly in this new scheme and was replaced by a new scientific language for motivation and volition.

One of the earliest and most controversial proponents of the new scientific views was Julien Offray de La Mettrie. His *Man a Machine* and *Man a Plant* (both 1747) offered a radical new conception of intelligence, which he attributed in varying degree to all animals. According to La Mettrie, a reliable measure of an organism's intelligence is the number of needs it must satisfy and the complexity of those needs. As La Mettrie was well aware, this complexity reflects the interconnectedness of all living organisms. Amoebas are among the simplest organisms, yet their survival depends on the satisfaction of nontrivial needs, such as thermal equilibrium, nourishment, and navigation.[4] Plants must compensate their weight and orientation by regulating their water intake and their absorption of energy from sunlight, which requires tracking the position of the sun.[5] Plants, like all forms of life on Earth, must behave in a time-sensitive fashion, dependent on their circadian rhythms, with critical implications for their behavior.[6] An organism's number and variety of needs quickly increase with the addition of locomotion, predation, and social pressures.

All organisms, according to La Mettrie, must satisfy needs, but not all have a mind. To have a mind, an organism's needs must be satisfied autonomously or by itself, though minds are also mechanical components of the physical universe. This assertion was important to La Mettrie because he wished to reconcile his mechanical view of the mind with the philosophical theories of his contemporaries, a task that remains central to physicalist and reductive views of mental processes. It was especially difficult to adapt conceptions of intelligence (as well as free will and motivation) to the materialistic view of the mind. In his endorsement to *Man a Machine* and *Man a Plant*, Noam Chomsky writes the following about La Mettrie's approach: "La Mettrie's inestimable contribution was to draw the natural conclusions from Cartesian physiology and Newton's radical revisions of traditional mechanics: that thought is a property of organized matter, on a par with electricity, the faculty of motion, and others—that mind is to be studied in the framework of the emerging scientific naturalism of the day."[7]

In spite of his strong commitment to naturalism, La Mettrie considered autonomous or agential need satisfaction as the key feature that demarcates the boundary between mindless and mindful life. A mechanical human is not merely a physical automaton determined by the general laws of physics and chemistry. Informational needs are constitutive of an autonomous mind. They are also constitutive of agential freedom. The needs of intelligent creatures, in Gottfried Wilhelm Leibniz's terms, *incline without necessitating*. We are part of the physical world, but our cognitive needs make us autonomous. Plants are "necessitated" to satisfy their biological needs simply by growing. Perhaps all plant behavior is just growth.[8] Plants are capable of satisfying biological needs, but La Mettrie did not consider biological needs to require cognitive input. Intelligent agency requires more than mechanistic biology.

According to our post-Darwinian understanding of life, needs evolved as life transformed and diversified. Somehow humans developed needs that transcended the mechanical: our needs are representational, rational, emotional, aesthetic, moral, and spiritual. But there is nothing mystical or dualistic about this—these are cognitive needs, just like those of other animals, who, like us, thought La Mettrie, must satisfy a multiplicity of needs autonomously. The key insight La Mettrie proposed, and which subsequent philosophers of materialism did not take seriously enough, is that intelligence requires autonomy. The puzzle, then, is as follows: how can our attention be autonomous if it satisfies needs automatically? The mechanics of attention have enormous consequences for action and cognition and, through joint attention, for culture and politics. This question is fundamental to determining whether and how attention can be a source of good.

THE SEARCH FOR THE MATERIALITY OF ATTENTION

Over the following centuries, scientists continued to investigate the mind from a mechanistic and materialistic perspective. Attention was quickly identified as essential to human intelligence, but scientists approached it from the perspective of various disciplines, which can be grouped into three categories: reflexology, behaviorism, and computational psychology.[9]

Reflexologists were interested in the physicochemical correlates of the reactions underlying behavior, understood in terms of the electrophysiology of tissues and nerves. Although their goals were not incompatible with those of behaviorists (in fact, both disciplines' goals were mutually supportive at various points), their emphasis was on localizing concrete anatomical mechanisms of cognition. Among the reflexologists was Théodule-Armand Ribot, a philosophically oriented materialist who pioneered experimental psychology in France. He was concerned with identifying what he called the "mechanism," rather than the *effects*, of attention.[10] Ribot thought that psychologists were not taking the materiality of the mind seriously enough. Too much emphasis was being placed on what happens when humans pay attention, rather than on the material causes of attention. Attention is a highly selective, sensitive, and goal-oriented cognitive process—one of the most selective kinds of information processing in nature. Because of the findings of reflexologists and then neuroscientists, we now know that this highly selective information processing depends on unique neuroanatomical structures and functions.

But this sophisticated scientific understanding of the mechanism of attention cannot explain the significant relationship between attention and autonomy or the implications of attention routines for the flourishing and well-being of the self and society. While Ribot rightly described automatic or spontaneous attention as primitive and "natural," as opposed to voluntary and cultural, his and subsequent mechanical or "motoric" accounts of attention are insufficient to capture the normative and personal aspects of attention. Such accounts, however, explained a great deal of why the mechanism of attention is a source of good. Automatic, involuntary, and largely unconscious kinds of attention satisfy many cognitive needs. Consider the problem of parsing perceptual scenes into objects and properties.[11] Solving this problem for olfaction is quite different from solving it for audition, both in terms of spatiotemporal structures and the nature of the stimuli. We can voluntarily move our attention from one modality to the other, but the features of olfactory and auditory scenes can powerfully trigger our attention automatically. The features that integrate the objects of our attention are parsed into meaningful units—they are never fully unorganized or unfamiliar—and this all happens automatically.

The behaviorists constituted a diverse group focused mostly on the mechanics of learning and the interpretation of behavior. They studied the patterns of arrays of behavior, the probability and organization of behavioral responses, and the flexibility or adaptiveness of learning new behaviors—a scientific categorization of the repertoire of concrete actions that help animals and humans solve various problems. These kinds of data could be put to further scientific scrutiny by being subjected to mathematical or statistical analysis. The behaviorists recognized that learning, training, and creating required skill and habituation. Their proposal was that the mind can be defined in terms of the action patterns involved in creating a behavior.

Along with Ribot and many other researchers across Europe, psychologists in Russia, such as N. N. Lange and V. M. Bekhterev, developed "motoric" approaches, which construed attention as a form of dexterity that frames all skilled perception, action, and cognition—a virtuous type of mental reflex.[12] Motoric theories had the advantage of providing a materialistic solution to questions of the causes of attention, rather than its multifarious effects. Attention could be *measured* by reactions, as well as by "action sets" and their anatomical counterparts. But this approach was also too reductive. Ironically, it was N. F. Dobrynin, a leading psychologist of the early Soviet period, who drew attention to the faults of this materialistic conception:

> All these theories did not take into account, however, that although attention is actually accompanied by certain adaptation movements, it is in no sense reducible to them. Certainly, if the spectator turns away from the stage, covers his eyes and plugs his ears, he will not be able to be attentive to what is occurring on the stage. But, in order to observe the stage and listen to what is

being said there, he must divert himself from everything else and direct perception to what is going on on the stage. It is possible to look and not see, as it were, to listen and not hear. Attention consists of seeing what you are looking at. The concept of attention can include within itself set or adjustment [patterns of skilled behavior while performing a task] in such an understanding of it, but it is considerably broader than this concept, and the crux of the matter is that its essence lies not so much in adaptation movements as in the selective character of the psychological activity involved.[13]

Similar to the frequently quoted observations of William James,[14] Dobrynin's characterization of the essence of attention was as a kind of cognitive selection for focusing on and dealing with the world: the "taking possession by the mind" of one salient option, in a vivid way opposed to distraction, out of many alternatives simultaneously presented to the mind. Elements of James's and Dobrynin's definitions can be found in contemporary views of attention that emphasize its parsing of information into foreground and background, the integration and unison of information, and the relevance of integrated attention in working memory, as well as in relation to action guidance, including mental action and inference.[15]

Dobrynin also emphasized the social dimension of the attentive mind. Habits of attention and attentiveness itself are culturally mediated and socially fostered, controlled, and regulated. These habits orient us toward memories, groups, educational trajectories, and preferences. We trust by paying attention to what is socially familiar and salient, and we entrust our attention to educators, politicians, and other sources of information. But there is tension between epistemic and moral trust and between the reflective and the automatic aspects of attention. With too many options and a flood of misinformation, whom we choose to trust with our attention has become a complex problem. In this context, ignoring what is not salient or familiar is essential, but doing so also creates unreliably biased echo chambers. By being selective, we become uncurious and indifferent.

Both reflexologists and behaviorists provided a powerful way to match biomechanical markers with patterns of behavior (even though they also identified flaws in their approaches), their work eventually leading to the field of contemporary neuroscience. The computer revolution in the mid-twentieth century gave rise to a third group of psychologists and linguists, beginning with the work of Alan Turing, Noam Chomsky, and David C. Marr. This group interpreted mechanism and function in terms of computational complexity and information theory. Neuroanatomy and behavior were thus understood as the material instantiation of mental information processing, from input to output. Accordingly, the intelligent mind was considered to be independent from the neural "hardware." Remarkably, despite their radically different approaches, these three camps—reflexology, behaviorism, and computational psychology—converged on attention as the key mechanism of intelligence and cognition.

The computational perspective in psychology and the reinterpretation of reflex theory in behavioral neuroscience were crucial for the eventual digitalization, localization, and, unfortunately, surveillance of our attention capacities in terms of the data structures that now drive the attention economy (and the scientific understanding of the mind). While the approaches of the reflexologists, behaviorists, and computational psychologists converged on attention, the computational approach was critical for uncovering the selective functions of attention. Too much information is available, and only a select set of informative items can be processed properly. Attention solves this problem by *inhibiting* irrelevant information. But this solution comes at a cost. Inhibition is essentially a form of bias toward certain types of information. Since attention operates by automatically inhibiting what is not relevant, it can be manipulated and its function distorted.

AUTONOMY AND ATTENTION

The dichotomy of selection and inhibition presents a puzzle. Attention as mechanism provides the automatic and selective biasing required to successfully navigate a vast sea of information, and, because of its automaticity, attention is "reflex-like." When our attention inhibits irrelevant or unreliable information, it is a source of knowledge. But when attention inhibits valuable information, it is detrimental. In both cases, attention seems to be out of our reflective control. Thus, the puzzle is how to train attention, given its algorithmic or mechanical nature. How are we to determine what is good or bad to attend to if attention is reflex-like?

To answer these questions, let us consider how two major theories conceive of attention as feature or object based.[16] Both theories try to solve the "binding problem" regarding how attention manages to integrate environmental information into objects, properties, foreground, background, and completely unified perceptual scenes. Automatic sensitivity and "good biases" are essential for knowledge acquisition. If attention is biased toward objects, then the problem is solved by the segmentation of objects from perceptual space. If it is biased toward features, then a map-like informational structure integrates features selected independently into objects with properties. This problem is so basic that without a solution, virtually no cognitive or representational need can be properly satisfied.

Perceptual attention solves the binding problem within sense modalities and through cross-modal integration. The smell of the ocean is immediately integrated with the sound of the waves. We can touch the water and feel the strength of a receding wave, and our visual system confirms this information: the wave looks strong. In this way, sensorial attention integrates evidence; therefore, it provides a normative basis for the justification of our beliefs. What we touch confirms what we see and hear. This selective "evidential cross-talk" is based on the reflex-like biases of attention toward select objects and properties, which actively inhibit other potential objects of attention, which then recede into the perceptual background (e.g., of voices, buildings, or people).

Attentive integration has a virtuous effect because it allows the attentive agent to navigate the world effectively. She is a good source of information, and others can rely on her. Her representational needs are satisfied because of how her attention routines solve complex binding problems. However, it is not clear that this kind of need satisfaction provides a perspective on the world that the agent could call her own. An autonomous perspective on the world cannot merely be the effect of a set of mechanisms, no matter how reliable. Thus, while computational and neural approaches to attention can explain attentional virtues for epistemic agency, they are inadequate to explain attentive virtue in general. On a social scale, if attention is indeed a source of civilization, as Ribot remarked, it must be firmly based on moral and aesthetic attentional capacities, which cannot be the mere effect of a set of informational mechanisms.

VIRTUOUS AND VICIOUS SENSITIVITY: THE PERSONAL AND CULTURAL EFFECTS OF ATTENTION

Perhaps virtue can be "built in" automatically or by design? The computational approach provided a useful insight by classifying attentional processes as either "bottom-up" or "top-down." This distinction categorizes attentional influences that are highly individualized, or *personalized*. Top-down attentional processes can embed routines for long-term goals, integrating them into a preference-based perspective that inhibits and suppresses immediate reactions or rewards. In this way, the selection processes of lower-level perceptual attention become deeply attuned to personal preferences and values. This kind of bias makes attention insensitive to information that is irrelevant to long-term action planning. When it is adequate, such insensitivity focuses limited cognitive resources on valuable goals that are not immediately imposed by the environment. It also makes possible joint attention to abstract contents, such as contracts and plans, which organize large-scale cooperation and collective action.[17]

Human attention is profoundly influenced by concepts and inferences. The perceptual landscape of attention routines is enriched by inferentially related informative structures, which are much more powerful sources of inhibition and selection.[18] Humans can pay attention for hours to a sequence of spoken or written sentences (as when we read). Indeed, we spend much of our time paying attention to language at the expense of vast amounts of other types of information available to us. We pay attention to the utterances of others, to written language, and to inner speech, which dominates most of our planning. Such a conceptual perspective constitutes a type of dexterity that can "time travel" to the past and future. These attentive biases can also be unconscious and determine what we see.[19] They allow us to pay joint attention to speech acts, generating a new background and foreground of information that includes speakers' intentions.[20]

An automatically structured perspective on the world underlies our linguistic and perceptual capacities. But rather than being highly individualistic and personal, the attention routines involved in creating this perspective are designed to satisfy *collective* epistemic needs, such as effective linguistic communication and large-scale cooperation. Sigmund Freud introduced the notion that the unconscious exerts a powerful influence on the mind based on the effects of culture on our implicit interests and desires. The attention we pay to the concrete and the abstract, to the perceptual and the social, and to the sensorial and symbolic operates immediately in our minds and shapes the structure of our motivations and desires.

The integration of top-down and bottom-up attention routines creates a scaffolding for joint attention and joint projects, which under certain circumstances can become addictive or destructive. With social media, these capacities have been industrialized and exploited on a global scale. The virtuous epistemic features of automatic attention geared toward social cooperation have become sources of attentional enfeeblement. It might be time to reconsider traditional forms of attention habituation that emphasize emotional and affective needs, rather than selfish and pragmatic needs. A healthy perspective on the world must include the proper satisfaction of moral and aesthetic needs.

ATTENTION FOR AFFECTIVE AND EMOTIONAL NEEDS

Our attentive "urges" may come into conflict. The era of digitalization has complicated this dynamic. On social media platforms, language has become empathically detached and quasi-anonymous, favoring abusive and obsessive kinds of competitive attention.[21] Animals without the complexity of human language satisfy communicational needs, but these needs and their associated urges are not as dominant as they are in humans. An extraordinary consequence of thinking linguistically on human cognition is that we became repositories of very productive and flexible kinds of classification capacities.[22] This development certainly was beneficial for our species, but it came at a cost: it generated new sources of constant stress and prevented other forms of attention from flourishing.

Evidentially grounded attention is the basis of trust in our assertions and communal knowledge. Moral and aesthetic virtues of attention are based on values that enrich our lives beyond knowledge and communication, such as the value of living a good life and spending time contemplating beautiful things. These virtues of attention provide a perspective on the world that can differ radically from the epistemic perspective. Both perspectives are essential, and they interact in various ways. The categorization-based perspective, however, has become dominant. The work of Nomy Arpaly illuminates how these perspectives may come into conflict.[23] Arpaly examines cases in which traditional theories that appeal to belief-based conviction are inadequate to explain good moral behavior. One of her examples comes from Mark Twain's *Adventures of Huckleberry Finn*. As Arpaly explains, Huckleberry Finn helps Jim, and this action is morally good, in spite of the fact that Finn *believes* that helping Jim is wrong. In this case, according to Arpaly, being akratic (incoherent or unwilling to act according to what one believes) is *good*. Finn's belief dictates that he not help Jim, but he suppresses the urge to act according to his belief because of the morally salient need to help Jim.

What is relevant to Finn is not his belief regarding the wrongness of helping Jim but rather his awareness of Jim's dignity.

Attending to our beliefs can provide a coherent perspective that serves as the foundation for epistemic autonomy. But in Finn's case, epistemic incoherence is morally virtuous. Suppressing epistemic attention to belief can also enable a moral perspective on the world—a perspective that provides a *different* kind of autonomy. Morally and aesthetically guided attention routines may seem automatic, too, but they are neither purely representational nor purely mechanistic. Consider the property that Finn attends to: human dignity. This feature (if one can call it a "feature") is too complex and at the same time too fundamental to be explained in terms of mechanisms and their effects. If all we have in mind are truth, conceptualization, and evidence, we become enfeebled moral agents, incapable of surrendering our belief-based perspective to the goodness and beauty of the world. We must balance our epistemic and moral cognitive needs. In the current context of our digitalized and commercialized world, this difficulty takes on global proportions.

UNFAMILIARITY: DIGITALIZED AND ENFEEBLED ATTENTION

Attention affects everything we do, so we should consider the extent to which our attention is under our control. Our evolutionarily established sensitivity to danger and social bonding can be used to influence and manipulate us.[24] These powerful sources of attentive salience have now been exploited through visually attractive platforms, through which our attention is captivated. A social paradox has emerged: the faculty that made us intelligent and autonomous is now making us dependent and vulnerable. We spend an unhealthy amount of time looking at screens, wanting and competitively seeking attention from others. In this environment, our attention is guided by industrial forces that turn all the activities on our screens into data and behavioral profiles. Our mental lives inhabit a virtual world that parasitically uses, monitors, measures, and exploits our attention with unprecedented success—a form of commercialization that Shoshana Zuboff aptly labels "surveillance capitalism."[25]

Given that autonomy matters in our lives, the question is how to tend to our autonomy—or different kinds of attentive *autonomies*. What should we be paying attention to in order to maintain a free and healthy perspective on the world? This is the key question concerning the virtues and norms of attention. We need to let our attention recover so that it can regain the intensity that experiences can afford when we fully devote our attention to them. This is a process of *familiarization* with the world. It is a slowing-down of our mental routines and the 24-7 clock. The assertiveness of the autonomous self who seeks to become her own ruler with a shield made of beliefs can be detrimental. Attention must stretch and reach out to what is valuable for it to be free and genuinely anchored in the world, not for commercial reasons but for its flourishing and engagement with others. Interconnection is not the same as genuine communication and engagement.

The commercial exploitation of our attention is producing an illusion of access with respect to who we are. In fact, we are becoming unfamiliar with ourselves and with one another. The problem is deeper than the threat posed by social media. A few decades before this technology was developed, our way of valuing one another had become influenced by where we fell on a scale of "positional" goods, goods considered valuable because of the reputational or financial prestige they bring. These artificial goods created fake personalities defined in terms of cars, houses, and boxed profiles that made accurate behavioral prediction possible. Our attention had been commodified and gamified before it became digitalized. Status-tracking attention is a fundamental part of our primitive instincts. We naturally seek social recognition and try to identify trustworthy authority figures. The market-based exploitation of our solidarity is a perversion of our natural trust in one another. The false comfort that comes with the commodification of attention has made us more mechanical and predictable.

It is imperative that we prevent the further erosion of our attentional habitats. We must create a new atmosphere for our attentive trust, away from the screens that dominate our minds through the mechanical aspects of attention. Human rationality unifies various forms of intelligence with emotions.[26] But drawing a sharp distinction between the epistemic and moral perspectives shows that balance matters for developing healthy kinds of attention—and we need balance more now than ever before.

NOTES

1. N. F. Dobrynin, "Basic Problems in the Psychology of Attention," in *Psychological Science in the USSR*, vol. 1, ed. B. G. Anan'yev, (Washington, DC: U.S. Department of Commerce, Clearinghouse for Federal Scientific and Technical Information, 1966), 274–91.
2. Carlos Montemayor and Harry H. Haladjian, *Consciousness, Attention, and Conscious Attention* (Cambridge, MA: MIT Press, 2015); Abrol Fairweather and Carlos Montemayor, *Knowledge, Dexterity, and Attention: A Theory of Epistemic Agency* (Cambridge: Cambridge University Press, 2017); Carlos Montemayor, "Inferential Integrity and Attention," *Frontiers in Psychology* 10 (2019): 2580, https://doi.org/10.3389/fpsyg.2019.02580.
3. By *virtue*, I mean the stable character traits and abilities of agents that make them trustworthy, reliable, and good (morally or epistemically). Virtues are a kind of excellence in action and performance.

4. Lucy Hicks, "Watch Amoebas Solve a Microscopic Version of London's Hampton Court Maze," *Science*, August 27, 2020, https://doi.org/10.1126/science.abe5316.
5. Andrew J. Millar, "A Suite of Photoreceptors Entrains the Plant Circadian Clock," *Journal of Biological Rhythms* 18, no. 3 (2003): 217–26.
6. Charles R. Gallistel, *The Organization of Learning* (Cambridge, MA: MIT Press, 1990); Carlos Montemayor, *Minding Time: A Philosophical and Theoretical Approach to the Psychology of Time* (Leiden: Brill, 2013).
7. Noam Chomsky, Endorsement for *Man a Machine and Man a Plant* by La Mettrie, trans. Richard Watson and Maya Rybalka (Indianapolis: Hackett, 1994).
8. For criticism, see Carrie Figdor, *Pieces of Mind: The Proper Domain of Psychological Predicates* (Oxford: Oxford University Press, 2018); Miguel Segundo-Ortin and Paco Calvo, "Are Plants Cognitive? A Reply to Adams," *Studies in History and Philosophy of Science* 73 (2019): 64–71.
9. This tripartite distinction is arbitrary, and it makes more sense in the context of the United States than Europe. Research in the United States was dominated by the rivalry between the behaviorists, led by B. F. Skinner, and the computational cognitive psychologists, inspired by the work of Alan Turing and Noam Chomsky. In fact, historically, the work of many groups overlapped. A study demonstrating the interconnected efforts of anatomists, reflexologists, and therapists is provided in Katja Guenther, *Localization and Its Discontents: A Genealogy of Psychoanalysis and the Neuro Disciplines* (Chicago: University of Chicago Press, 2015). In the Soviet Union, although research was heavily influenced by Ivan Pavlov, various approaches went beyond a mechanical explanation and emphasized society and culture.
10. Théodule-Armand Ribot, *Psychologie de l'attention* (Paris: Germer Baillière, 1889).
11. Later in the chapter, I discuss how perceptual attention solves the "binding problem."
12. Dobrynin, "Basic Problems," 274-275. Dobrynin was a member of the so-called activity theorists. See Yuri Dormashev and Evgeny N. Osin's chapter in Brian Bruya, *Effortless Attention: A New Perspective in the Cognitive Science of Attention and Action* (Cambridge, MA: MIT Press, 2010, 287-334).

13. Dobrynin, "Basic Problems," 275.
14. William James, *The Principles of Psychology* (New York: Henry Holt, 1890).
15. On parsing information, see Sebastian Watzl, *Structuring Mind: The Nature of Attention and How It Shapes Consciousness* (Oxford: Oxford University Press, 2017); on integration and unison, see Christopher Mole, *Attention Is Cognitive Unison: An Essay in Philosophical Psychology* (New York: Oxford University Press, 2011); on integrated attention in working memory, see Jesse J. Prinz, *The Conscious Brain: How Attention Engenders Experience* (New York: Oxford University Press, 2012); in relation to action guidance, see Wayne Wu, "Attention as Selection for Action," in *Attention: Philosophical and Psychological Essays*, ed. Christopher Mole, Declan Smithies, and Wayne Wu (New York: Oxford University Press, 2011), 97–116; on mental action and inference, see Montemayor, "Inferential Integrity."
16. See, for instance, Daniel Kahneman, Anne Treisman, and Brian J. Gibbs, "The Reviewing of Object Files: Object-Specific Integration of Information," *Cognitive Psychology* 24, no. 2 (1992): 175–219; Anne Treisman and Garry Gelade, "A Feature-Integration Theory of Attention," *Cognitive Psychology* 12, no. 1 (1980): 97–136; Zenon W. Pylyshyn, "Visual Indexes, Preconceptual Objects, and Situated Vision," *Cognition* 80, no. 1–2 (2001): 127–58.
17. Herbert H. Clark, *Using Language* (Cambridge: Cambridge University Press, 1996); Dan Sperber, *Explaining Culture: A Naturalistic Approach* (Oxford: Blackwell, 1996).
18. Montemayor, "Inferential Integrity."
19. Susanna Siegel, *The Rationality of Perception* (New York: Oxford University Press, 2017), 147–95.
20. Clark, *Using Language*; Fairweather and Montemayor, *Knowledge, Dexterity, and Attention*.
21. Robert Sapolsky, "To Understand Facebook, Study Capgras Syndrome: This Mental Disorder Gives Us a Unique Insight Into the Digital Age," *Nautilus*, October 27, 2016, https://nautil.us/to-understand-facebook-study-capgras-syndrome-236173/.
22. Robert C. Berwick and Noam Chomsky, *Why Only Us: Language and Evolution* (Cambridge, MA: MIT Press, 2016).

23. Nomy Arpaly, *Unprincipled Virtue: An Inquiry Into Moral Agency* (New York: Oxford University Press, 2002).
24. Frans de Waal, *Mama's Last Hug: Animal Emotions and What They Tell Us About Ourselves* (New York: Norton, 2019).
25. Shoshana Zuboff, *The Age of Surveillance Capitalism: The Fight for a Human Future at the New Frontier of Power* (New York: Public Affairs, 2019).
26. Luiz Pessoa, "On the Relationship Between Emotion and Cognition," *Nature Reviews Neuroscience* 9, no. 2 (2008): 148–58.

6

ATTENTION, TECHNOLOGY, AND CREATIVITY

CAROLYN DICEY JENNINGS AND SHADAB TABATABAEIAN

A SCENE OF ATTENTION: THE BREAK

Imagine a scientist who is working on an experiment, reading articles and jotting down notes for the framework, when she reaches an impasse. She realizes that one part of the experiment won't work. She backs her chair away from the desk, frustrated, deciding to take a break. She scrolls through Twitter on her phone, seeing posts from her friends and colleagues—some professional, some less so. She walks to the kitchen and looks out the window as she pours herself a glass of water, watching tree branches sway in the wind. As she walks back to her office, she has an idea about how to move forward—a creative insight. Did her break contribute to this insight? Did Twitter? In this chapter, we use findings from psychology and neuroscience to help determine the impact of recent digital technologies, such as smartphones, social media, and online games, on attention and creativity. We start with some background information on the scientific conception of attention and creativity before discussing findings on the use of recent digital technologies and what they mean for us as autonomous, creative beings.

A SHORT PRIMER ON THE SCIENTIFIC CONCEPTS OF ATTENTION AND CREATIVITY

Attention is one of the most basic mental functions. In the brain, it allows for the selection of preferred neural signals by increasing their strength through neural feedback, thus diminishing the relative strength of other neural signals. It is attention that allows the scientist, for example, to focus on her experimental design while people chat nearby, by emphasizing her thoughts at the expense of those conversations.

Attention comes in many forms, but most relevant to this chapter are "top-down" and "bottom-up" attention. These concepts are based on the much older division between voluntary and involuntary attention, or the difference between using attention to further a goal or interest, such as when working on an experiment, and one's attention being occupied by a distracting stimulus, such as people chatting nearby.

The difference between voluntary and involuntary attention is measured in behavior through "endogenous" and "exogenous" cues.[1] Whereas endogenous cues are symbolic and direct attention toward a target (e.g., an arrow), exogenous cues are salient and capture attention at the location of the target (e.g., a flash). Exploring this difference with brain imaging allowed researchers to distinguish the brain networks responsible for top-down attention, associated with endogenous cues, from those responsible for bottom-up attention, associated with exogenous cues.

Importantly, these top-down and bottom-up networks are part of a larger system of attention.[2] One can envision this as a system of resource distribution, with top-down attention favoring one's current goals and bottom-up attention favoring salient stimuli, with one's focus landing on one or the other depending on the strength of each.

Top-down attention is key to mental control and autonomy: the ability to act in accordance with one's interests and values. Whereas bottom-up attention occurs for salient stimuli, top-down attention occurs when one's current task is maintained despite the presence of those stimuli. Mental control and autonomy are based in this ability to exert control over mental resources—and it is this control that some fear is under threat from the use of recent digital technologies.

A separate issue often brought up in discussions of these technologies is whether creativity is helped or hindered by their use. Creativity requires both novelty and utility; according to Berys Gaut, "There is a broad consensus that creativity is the capacity to produce things that are original and valuable."[3] Both conditions (novelty/originality and utility/value) can be satisfied at either the subjective or objective level, and need not be associated with monetary or other social gains. The creative insight of the scientist might be novel to her but not to others; a new scientific hypothesis could be useful because it furthers understanding for the scientist or for the field at large. Ruled out are cases of useless novelty, such as gluing a hammer to a nail.

Psychologists have found that creative ideas are formed through a dual process of generation and evaluation.[4] During the generation process, many ideas are explored, whereas during the evaluation process, the ideas are assessed and further elaborated, leading to the selection of those that are most novel and useful. The scientist, for example, might first explore the space of possible solutions and then apply them to determine the optimal solution. The brain does not have a specialized area for creative cognition; rather, creativity arises from interactions among multiple cognitive mechanisms, such as memory and executive function.

Attention and creativity may at first seem opposed to each other: attention is associated with control and constraint, whereas creativity is associated with novelty and growth. Yet the connection between them is more complex. Some argue that creativity occurs in the absence of attention, but others argue that different forms of attention contribute to different forms of creativity.[5] We argue, instead, that the condition of utility, engendered by the evaluation phase of creativity, requires a role for attention. Thus, if recent digital technologies reduce our capacity for attention, they are also likely to negatively affect our creativity. We will come back to these issues after we first examine the evidence that recent digital technologies undermine top-down attention.

THE IMPACT OF RECENT DIGITAL TECHNOLOGIES ON ATTENTION

Recent digital technologies target attention. As we will discuss, they have been designed to manipulate attention by covertly capturing and engaging it; they are interactive in very short time spans, increasing their ability to capture attention; and they are adaptive, in that they can easily adjust to better manipulate attention. These facts are concerning given the link between attention and autonomy described earlier: "We may justifiably worry about whether social media are undermining our ability to shape our own lives by making us less able to focus on our goals and more likely to chase after immediate diversions."[6] Indeed, our review of the evidence found that recent digital technologies appear to have a lasting impact on our ability to control attention in a top-down manner.

Common Design Elements of Recent Digital Technologies

A common design element used to manipulate attention in recent digital technologies is *intermittent variable rewards*, developed following their use in slot machines.[7] These are rewards that vary both in rate (how often they occur) and magnitude (how large they are). The purpose of intermittent variable rewards is to keep the user's attention on the product by preventing them from being able to predict when they will receive the reward and how valuable the reward will be. Social feedback is naturally intermittent and variable, but this effect is amplified by social media when it, for example, adjusts the timing of notifications received by the user so that they arrive in less predictable batches. This and other techniques are used to capture the user's attention at the expense of other activities without the user's explicit awareness.[8]

While many technologies have used intermittent variable rewards, recent digital technologies are more interactive than earlier technologies, making them much more successful at capturing attention. Take the difference between watching an advertisement on your television versus your smartphone: your attention may be captured by each, but only with your smartphone does the advertisement change as a result of your attention (e.g., through hovers and clicks). If you tap an advertisement on your smartphone, for example, you will likely be led to more information about the product being advertised.

What's more, the scale of interaction is much greater than with earlier technologies: within the same amount of time, interactions occur with far more information, individuals, and corresponding spatial coverage. These factors make an interaction more salient than one involving less information, fewer individuals, or less spatial coverage. At the same time, the interaction is much faster, providing near immediate feedback to the user. Immediacy is important because even somewhat delayed feedback requires greater cognitive resources for attention to be maintained, making continued interaction less likely.

Finally, recent digital technologies are adaptive. They adjust to the user's hovering and clicking behavior and even the user's personality and mood. This adaptability makes these technologies particularly well suited to manipulating our attention, sometimes using our vulnerabilities against us.[9]

The Measurable Impact of Recent Digital Technologies on Attention

Given the design features of recent digital technologies, one might question the extent to which these technologies are *effective* at manipulating attention. They might be considered effective if they capture attention against the wishes of the user, or at least against the better judgment of the user. Take, for example, the off-task use of smartphones by students in the classroom. It has been demonstrated that such use has a negative impact on academic performance, yet this behavior is widespread. Assuming that it is in students' best interest not to use their smartphone in class, this evidence appears to demonstrate that smartphones capture students' attention against their better judgment.

Is this evidence of a short- or long-term impact on attention? It could be both. If we take a short-term perspective, we treat the student's self-control as fixed and thus a confounding factor (i.e., low self-control causes both in-class smartphone use and lower grades). In one study taking this perspective, accounting for personal and course characteristics revealed a reduced effect of smartphone use on grades.[10] Yet if we take a long-term perspective, self-control may not be a confounding factor, since it might have been reduced by previous smartphone use.

One review found evidence that recent digital technologies reduce attention capacity in the short term but limited evidence for a longer-term impact.[11] To provide an example of a short-term effect on attention, the authors described a study in which mere exposure to smartphone notifications detracted from performance in an ongoing task. To explore long-term effects, the authors examined studies on media multitasking looking at the correspondence between a person's level of multitasking (e.g., how often they look at their phone while watching television) and their ability to direct and control their attention; in this instance, the evidence is mixed. However, a more recent review of media multitasking found that "most studies to date report negative effects of media multitasking on measures of attention," citing impacts on both brain and behavior.[12]

While the direct evidence for a long-term impact on attention is limited, the potential for such an impact is supported by evidence on addiction. Recall that recent digital technologies can be considered effective at manipulating attention if they capture attention against the wishes of the user—the hallmark of addiction. Addiction to recent digital technologies has become about as prevalent as alcohol addiction in the United States, with smartphone addiction seemingly replacing drug and alcohol addiction in younger generations.[13] While this extreme case of long-term impact is limited to a subset of the population, it is reasonable to suppose that the long-term impact on attention experienced by this group is also experienced, to a lesser degree, by other users. That is, those with internet addiction disorder may be *unable* to control their use of recent digital technologies, whereas others are simply *less able* to control it.

Given the evidence discussed, we think there is reason to suppose that recent digital technologies have a long-term negative impact on our ability to direct and control attention. Our basic reasoning is as follows:

1) The use of recent digital technologies has been demonstrated to temporarily reduce one's ability to sustain attention.
2) Brain plasticity allows the short-term effects of an activity to translate into long-term tendencies.
3) Both brain and behavioral differences in attention have been demonstrated in those who frequently use recent digital technologies, in keeping with points 1 and 2.

This basic argument has been made by other authors: "the assumption of long-term effects of smartphone usage on cognitive functions based on findings from related fields . . . and the knowledge that brain structures can be altered with prolonged exposure to novel environments."[14] In other words, we know that changes to brain and behavior in the short term can translate into long-term changes. Therefore, it is reasonable to suppose that the short-term effects of recent digital technologies on attention that have already been established will be found to be at least partially responsible for long-term brain and behavioral differences discovered in those who frequently use such technologies—namely, the reduced capacity for sustaining attention toward a current goal or interest in the face of distractions.

Modeling and Evaluating the Impact of Recent Digital Technologies on Attention

Assuming that recent digital technologies have an impact on attention, how might we model and evaluate that impact? In terms of modeling, there appear to be at least two effects relevant to our purposes: making the technology in question more likely to capture bottom-up attention and making top-down attention less effective.

On the first effect, we have already seen how recent digital technologies are highly effective at capturing bottom-up attention: they use methods that make engaging with these technologies seem more valuable to the user. We call this "artificial salience." Recall, for instance, the widespread adoption in recent digital technologies of intermittent variable rewards, known to cause addictive behavior. Another common feature of these technologies is that they enable social connection and feedback, one of our greatest motivators. Combining the two features is a powerful recipe for capturing attention.

As to making top-down attention less effective, this occurs when recent digital technologies bias the user toward bottom-up, rather than top-down, attention in the long term. We call this "exploration bias." One might see the balance between top-down and bottom-up attention as favoring top-down attention when it is important to persist in tasks ("exploiting" them) and as favoring bottom-up attention when it is important to continue exploring alternatives. However, Bermúdez describes how recent digital technologies encourage us to favor bottom-up attention regardless of the situation: "The internet enables us to process more things, but it simultaneously spreads our attention much more thinly to cover a wider area of content, and makes it continuously shift between tasks."[15] That is, by offering such a wide array of attractive options, these technologies shift our attentional *strategy* in favor of exploration over exploitation. We believe that over time, this strategy will become a trait, making the user less able to exert top-down attention in a sustained manner.

So, is the effect of recent digital technologies on attention good or bad? Answering this question isn't easy.

We do find artificial salience to be pernicious in some cases. That is, the effect is associated with addiction in a subset of the population, meaning that these users no longer have control over their behavior with respect to recent digital technologies. This loss of autonomy seems straightforwardly bad, just as in alcohol use disorder. Some have argued that the effect is also bad for the rest of the population: "By deliberately and covertly engineering our choice environments to steer our decision-making, online manipulation threatens our competency to deliberate about our options, form intentions about them, and act on the basis of those intentions. . . . Online manipulation thus harms us."[16] That is, insofar as these technologies purposefully undermine the user's ability to choose to occupy their time in another way and without the user's awareness, they are a threat to autonomy. In our view, much of the "badness" in this more general case depends on the extent to which these efforts are both deliberate and covert.

We are more equivocal about the increase in exploration bias. One might worry about the loss of top-down control in this case. That is, by altering the balance between top-down and bottom-up attention, recent digital technologies may be seen as threatening our ability to exert top-down control and thus as threatening our autonomy. Yet top-down control is never total, and for good reason. Being responsive to changes in our environment is important and sometimes more important than being able to persist in the task at hand. If our environment changed to one in which reactivity was more valuable than persistence, then it would be a good thing if our attention adapted to that environment.

What's more, exploration bias may be conducive to greater creativity. The popular press often claim that recent digital technologies improve creativity because they allow for a larger quantity of input: "The fact that everyone has access to a wealth and diversity of ideas and the means to actualize intent means that we all can be more creative."[17] If this is correct, being more receptive to such input may enable even greater creativity. However, studies on the connection between recent digital technologies and creativity provide mixed evidence.[18] In the next section, we briefly examine the connection between attention and creativity to help answer this question. We then conclude the chapter with some ideas on how to better explore the possibility of recent digital technologies enhancing creativity.

THE DEPENDENCE OF CREATIVITY ON ATTENTION

As described earlier, creativity is a "biphasic" phenomenon, in which one phase involves idea generation and the other idea evaluation. Idea generation benefits from having few constraints; that is, reduced top-down attention. Conversely, idea evaluation relies on increased top-down attention to assess ideas and then select and further elaborate the most appropriate ones, depending on the individual's goals and interests.[19] For example, in trying to solve a difficult problem, the scientist engages both top-down and bottom-up attention: she allows ideas to percolate up through bottom-up attention and exerts top-down attention to help constrain the ideas (as well as, perhaps, to inhibit irrelevant sensory stimuli). Of course, the extent of the contribution of top-down and bottom-up attention depends on the type, context, and attributes of the creative activity.

Counterintuitively, creativity is thus instantiated by leveraging *both* top-down and bottom-up attention. Neuroimaging studies of creativity corroborate this account. Creative cognition is supported by two large-scale brain networks: the default mode network (DMN) and the frontoparietal control network (FPN). The DMN is activated during episodic memory retrieval, self-reflection, thinking about the past, and planning for the future. The FPN is responsible for cognitive control and decision-making processes. The coupling of these two networks induces a unique state in which memory retrieval and imagination continue in the absence of irrelevant sensory stimuli, while, simultaneously, self-generated mental content is guided and constrained by top-down regulation to fulfill a goal. This coupling is unique since the DMN and FPN are typically characterized as opposing networks; that is, the activation of one usually suppresses the activation of the other.[20]

As we have described, creative cognition occurs when these two networks cooperate, supporting both idea generation (through the DMN and bottom-up attention) and idea evaluation (through the FPN and top-down attention). Interestingly, a similar neural pattern has been reported during certain mind-wandering episodes.[21] Many studies have found that mind-wandering enables creativity, particularly when an individual hits an impasse while working on a problem. During mind-wandering, the mind may be allowed to gravitate toward the unsolved problem, revisiting and exploring it with fewer constraints and from new angles.[22]

However, while mind-wandering is often associated with the absence of constraint, the cognitive process of enabling creativity must involve some degree of control and constraint. This is because creativity requires not only novelty but utility, as defined earlier. Thus, creativity requires top-down attention. In our view, creative cognition is associated with an increased coupling of the DMN and the FPN because creativity benefits from a closer connection between its two phases, with top-down attention playing a crucial role throughout the creative process.

Let's return to the example we started with. Recall that our scientist resolves an impasse after a break that involved both scrolling through Twitter and looking at the trees outside. We asked whether her break contributed to this insight and what the contribution of her Twitter engagement might be. We are now in a better position to answer these questions. On the first, since the type of thinking associated with creative cognition is unique, requiring the cooperation of brain networks that are typically opposed to one another, a break may be a good way to move into a state of creative cognition. On the second, while external stimuli may help trigger a creative idea, the cognitive state induced by the use of social media is generally opposed to creativity because it puts pressure on top-down attention and control. Looking at trees, on the other hand, has been widely cited as restoring attention during a break and so may serve as a better source of creative insight.[23] Looking at trees is one instance of mind-wandering that might lead to examining the problem from an unexplored angle. Mind-wandering is by definition a low-demand activity relative to, for example, continual interactions with distracting bottom-up stimuli in the form of phone notifications. We thus suspect that recent digital technologies are generally detrimental to both attention and creativity.

CONCLUSION AND FUTURE DIRECTIONS

We have established that recent digital technologies are disruptive to top-down attention and that disruptions of top-down attention are detrimental to creative cognition. Flurries of goal-irrelevant external stimuli designed to actively engage the user—flashing colors, noises, notifications, infinite scrolling—are likely to impede top-down attention and thus creative problem-solving. As we mentioned, it seems unlikely that scrolling through Twitter helped the scientist generate her creative insight; what is more likely is that the active and demanding nature of external stimuli originating from the Twitter app broke her chain of thought and distracted her from the problem she was working on.

Therefore, we find it unlikely that recent digital technologies typically foster creativity through changes in attention. While one hypothesis about creative cognition links it to reduced cognitive control, we have shown that the full story is much more complicated: reduced cognitive control may increase the number of ideas that one is presented with, but it keeps one from effectively making use of those ideas. Creativity relies on a balance between top-down and bottom-up attention. This balance allows for goal-directed memory retrieval, the generation of new mental representations, and then further elaboration and evaluation of those representations into useful outcomes. A constant stream of bottom-up stimuli, a chain of tweets for example, prevents this state of sustained attention, thus impairing processes conducive to creativity.

Recall the quote from earlier in the chapter: "The fact that everyone has access to a wealth and diversity of ideas *and the means to actualize intent* means that we all can be more creative." The ability to act on ideas is a crucial element of creative cognition, but this ability is undermined by recent digital technologies through the effects we have described. It seems unlikely that additional sensitivity to bottom-up stimuli will be enough to compensate for a loss in top-down control.

Yet the use of recent digital technologies may foster creativity in ways that we haven't considered here, and this notion would benefit from future work. First, it may be that the occasional use of these technologies fosters creativity by providing inspiration during sticking points. Key to this possibility will be the ability to recognize and make use of that inspiration, which will require that the impact of these technologies on cognitive control are short-lived. One might compare this situation with the effect of altered states, such as those induced by certain drugs, to foster creativity; breakthroughs may occur, but they will only be useful if they are harnessed when the altered state subsides.

Second, individual differences may cause certain people to become more creative through even habitual use of recent digital technologies. If we consider top-down and bottom-up control of attention to balance along a spectrum, many individuals may be skewed relative to the best fit with their environment. For those individuals, altering the balance of top-down and bottom-up attention through the use of recent digital technologies may bring them into better alignment with their environment and perhaps even allow for greater creativity. In other words, those with an excess of top-down control may benefit from a reduction in it, even if such a reduction wouldn't necessarily help people in general.

Finally, recent digital technologies may induce greater creativity through the content they provide. These technologies allow us greater access to books, academic articles, music, art, movies, and other content known to inspire creativity. Further, they allow us greater access to our social networks. Research on creativity in the business setting has found that individual creativity is important but that "the exposure to ideas from other team members and the use of creative problem-solving techniques may be at least as important."[24]

Thus, it may be that the benefits of recent digital technologies to creativity outweigh the deficits, at least for some individuals, but that harmful effects on attention remain. While we have argued that exploration bias may have advantages in certain environments, we find that artificial salience is harmful when it leads to addictive behavior and when it is designed to manipulate the user.

NOTES

1. Michael I. Posner, "Orienting of Attention," *Quarterly Journal of Experimental Psychology* 32, no. 1 (1980): 3–25.
2. Michael I. Posner and Steven E. Petersen, "The Attention System of the Human Brain," *Annual Review of Neuroscience* 13 (1990): 25–42; Steven E. Petersen and Michael I. Posner, "The Attention System of the Human Brain: 20 Years After," *Annual Review of Neuroscience* 35 (2012): 73–89.
3. Berys Gaut, "The Philosophy of Creativity," *Philosophy Compass* 5 (2010): 1034–46.
4. Oded M. Kleinmintz, Tal Ivancovsky, and Simone G. Shamay-Tsoory, "The Two-Fold Model of Creativity: The Neural Underpinnings of the Generation and Evaluation of Creative Ideas," *Current Opinion in Behavioral Sciences* 27 (2019): 131–38.
5. Darya L. Zabelina, "Attention and Creativity," in *The Cambridge Handbook of the Neuroscience of Creativity*, ed. Rex E. Jung and Oshin Vartanian (Cambridge: Cambridge University Press, 2018), 161–79.
6. Juan Pablo Bermúdez, "Social Media and Self-Control: The Vices and Virtues of Attention," in *Social Media and Your Brain: Web-Based Communication Is Changing How We Think and Express Ourselves*, ed. C. G. Prado (Santa Barbara, CA: Praeger, 2017), 57–74.
7. Natasha Dow Schüll, *Addiction by Design: Machine Gambling in Las Vegas* (Princeton, NJ: Princeton University Press, 2012).
8. Christian Montag et al., "Addictive Features of Social Media/Messenger Platforms and Freemium Games Against the Background of Psychological and Economic Theories," *International Journal of Environmental Research and Public Health* 16, no. 14 (2019): 2612.
9. Daniel Susser, Beate Roessler, and Helen Nissenbaum, "Technology, Autonomy, and Manipulation," *Internet Policy Review* 8, no. 2 (2019), https://doi.org/10.14763/2019.2.1410.

10. Andreas Bjerre-Nielsen et al., "The Negative Effect of Smartphone Use on Academic Performance May Be Overestimated: Evidence from a 2-Year Panel Study," *Psychological Science* 31, no. 11 (2020): 1351–62.
11. Henry H. Wilmer, Lauren E. Sherman, and Jason M. Chein, "Smartphones and Cognition: A Review of Research Exploring the Links Between Mobile Technology Habits and Cognitive Functioning," *Frontiers in Psychology* 8 (2017): 605.
12. Melina R. Uncapher and Anthony D. Wagner, "Minds and Brains of Media Multitaskers: Current Findings and Future Directions," *Proceedings of the National Academy of Sciences* 115, no. 40 (2018): 9889–96.
13. Jeff Cain, "It's Time to Confront Student Mental Health Issues Associated with Smartphones and Social Media," *American Journal of Pharmaceutical Education* 82, no. 7 (2018): 6862.
14. Magnus Liebherr et al., "Smartphones and Attention, Curse or Blessing? A Review on the Effects of Smartphone Usage on Attention, Inhibition, and Working Memory," *Computers in Human Behavior Reports* 1 (2020): 100005.
15. Liebherr et al., "Smartphones and Attention."
16. Susser, Roessler, and Nissenbaum, "Technology."
17. Greg Satell, "How Technology Enhances Creativity," *Forbes*, January 27, 2014, https://www.forbes.com/sites/gregsatell/2014/01/27/how-technology-enhances-creativity/?sh=3ff9ba6c3f50.
18. See, e.g., Marianna Sigala and Kalotina Chalkiti, "Knowledge Management, Social Media and Employee Creativity," *International Journal of Hospitality Management* 45 (2015): 44–58; Jana Kühnel et al., "Staying in Touch While at Work: Relationships Between Personal Social Media Use at Work and Work-Nonwork Balance and Creativity," *International Journal of Human Resource Management* 31, no. 10 (2020): 1235–61.
19. Evangelia G. Chrysikou, "The Costs and Benefits of Cognitive Control for Creativity," in *The Cambridge Handbook of the Neuroscience of Creativity*, ed. Rex E. Jung and Oshin Vartanian (Cambridge: Cambridge University Press, 2018), 195–210.
20. Roger E. Beaty et al., "Robust Prediction of Individual Creative Ability from Brain Functional Connectivity," *Proceedings of the National Academy of Sciences* 115, no. 5 (2018): 1087–92.
21. Kalina Christoff et al., "Experience Sampling During fMRI Reveals Default Network and Executive System Contributions to Mind

Wandering," *Proceedings of the National Academy of Sciences* 106, no. 21 (2009): 8719–24; Jonathan W. Schooler et al., "Meta-Awareness, Perceptual Decoupling and the Wandering Mind," *Trends in Cognitive Sciences* 15, no. 7 (2011): 319–26.

22. Shelly L. Gable, Elizabeth A. Hopper, and Jonathan W. Schooler, "When the Muses Strike: Creative Ideas of Physicists and Writers Routinely Occur During Mind Wandering," *Psychological Science* 30, no. 3 (2019): 396–404; Claire M. Zedelius and Jonathan W. Schooler, "Capturing the Dynamics of Creative Daydreaming," in *Creativity and the Wandering Mind*, ed. David D. Preiss, Diego Cosmelli, and James C. Kaufman (London: Academic, 2020), 55–72.

23. David Badre, "Tips from Neuroscience to Keep You Focused on Hard Tasks," *Nature*, March 15, 2021, https://www.nature.com/articles/d41586-021-00606-x; Andrea Faber Taylor and Frances E. Kuo, "Children with Attention Deficits Concentrate Better After Walk in the Park," *Journal of Attention Disorders* 12, no. 5 (2009): 402–409.

24. Monica J. Garfield et al., "Research Report: Modifying Paradigms—Individual Differences, Creativity Techniques, and Exposure to Ideas in Group Idea Generation," *Information Systems Research* 12, no. 3 (2001): 322–33.

7

ATTENDING TO ABSENCE, AND THE ROLE OF THE IMAGINATION

JONARDON GANERI

A SCENE OF ATTENTION

I am pacing around my living room. I am there because I have misplaced the novel I was reading a short while ago, and now I am on the hunt for it. A quick visual search is sufficient for me to establish that my book is not there. But what exactly does my knowledge that the book is absent consist in? And what are its grounds? The brilliant Bengali philosopher Krishnachandra Bhattacharyya (1875–1949) taught us that the grounds are ones of attention. In his early essay "Some Aspects of Negation," he says that a philosophical temperament is an attitude toward attention, a "mode of handling a given content."[1] He explains our scene of attention by introducing a new concept, the concept of negative attention:

> We know the absence of an object, say of a book on the table[,] by a faculty which is neither perception nor inference. It is not perception, for the absence gives no sensation; and it cannot be inference, for inference must be based on perception. The faculty however being there, it may be helped out by the perception or

inference. It is nearest to psychological introspection, though it knows objective nonexistence and not subjective existence merely. The nonexistence of a book on the table is an objective fact known by negative attention, defined by relation to the facts obtained by positive attention. Through this negative attention then, we also know a particular negation or absence of knowledge, know the want of a solution and therefore the solution itself. Negative introspective attention accordingly is the faculty that requires to be controlled.[2]

The faculty by means of which I know that the room is devoid of my book, it is claimed, is neither perception nor inference but attention, a claim that derives its plausibility from the fact that attention can be directed toward things that are not at the center of one's gaze and, indeed, toward things that one does not see at all. I might, for instance, be looking out the window but attending to a mathematical problem or a chess puzzle or to what I have to do later in the day or something that happened earlier. I can't *see* the absence of my book on the table; I can see only the table and its contents. And I don't *deduce* its absence from my failure to perceive it. I can, though, *attend* to the book's absence, a proprietary mode of attention made possible by what I do see and infer. Here, negative attention is attention to wants, lacks, losses, and lacunae—to the missing and the misplaced.

AESTHETIC PERCEPTION

By the time Bhattacharyya returns to this example, in his justly celebrated classic *The Subject as Freedom*, his discussion has acquired a new level of depth and sophistication. This is how he begins:

A person is looking for a book in a room but does not find it. He knows the book to be absent without being conscious of any empty look about the room and without, in fact, consciously referring the absence to the room at all. . . . What is known in the first instance is not the absence of the book but the book as absent. The book is not found and the room where it is not found is not perceived, at least immediately, to have any empty look owing to the absence. The book as absent is immediately known as a present objective circumstance that is neither remembered nor merely imagined. The missing of the imagined book is a characteristic experience, implying a feeling of the body not reaching it, which is interpreted as the objective fact of the book being absent. The knowledge of this fact of the book as absent is with the conscious imagination of the book as found being distinct from it and implies the feeling of the present fact being outside objective space altogether. Such knowledge also may, therefore, be called aesthetic or imaginative perception.[3]

The book is experienced as *missing* without, yet, any awareness of the room as *devoid*. To experience the book as missing is to have a feeling of the body not reaching it. My internal awareness of the position of my body, a kinesthetic sense of my body as arranged in space, makes me also aware of my immediate environment as being, as it were, "to hand," as within touching distance. The felt body creates within its environment a threshold of "to-handness," objects that fall within this threshold being "immediately known" as present and objects that fall outside immediately known as absent. My searched-for book is not within this felt penumbra of the body and so is experienced by me as missing. It is not *here*, in peripersonal space.[4]

The movement from experiencing the book as missing to knowing that it is absent involves the imagination. I *imagine* a contrary scenario in which the book is present, so that my knowledge of the book as absent consists in a contrast with an imagined scene. This is not a case of ordinary perceptual knowledge, because the object is not present in the perceived environment, but it is nevertheless sufficiently similar to ordinary perceptual knowledge to be plausibly described as a special case of perception, an "aesthetic or imaginative" perception as Bhattacharyya puts it. The idea of aesthetic or imaginative perception is that of perceptual knowledge that intrinsically involves the imagination. Imagination here is not a faculty of free fancy but the ability to manipulate images. Bhattacharyya introduces the idea by way of a slightly different example, in which one becomes aware that there used to be a tree in the field in which one is now standing, a tree that has now gone:

> The absence of the tree is known as a character of the locus, the perceived field where the tree stood. The tree may not be definitely remembered but if remembered it is recognised to be the specification of the absence that continues to be known, the place not ceasing to wear the bare look because of the definite memory. As the place is perceived, absence as a character of the place may also be claimed to be perceived. There is, however, a distinction between the sense in which absence is a character and that in which a quality like colour is a character of the place. The place in being perceived with the bareness or absence is, if not perceived, then at least imagined as what need not have the character, being presented as with a new look or, in other words, as distinct from what it might be. But to perceive the place with a colour is not necessarily to imagine that it might be without it. The perceived locus of absence being imagined in the very perception of it as without the absence, the absence is only a floating adjective that unlike colour is felt to be dissociated from the locus. . . .

> The perception [of the absence] may be called aesthetic or imaginative perception to distinguish it from ordinary perception.⁵

The field is perceived, and it is perceived as *bare*, as lacking in some as yet nonspecific way. I have not yet consciously remembered the tree but simply see the field as having a bare look about it. I then retrieve a memory of the field as it was, perhaps by journeying back to the field in episodic memory or mental time travel. I compare in imagination the contents of my retrieved memory with the presented scene, memory now supplying imagination with content to be manipulated. There is thus a distinction between seeing the field as bare, as having a bare look, and seeing it as green. Bareness is a "floating adjective" in the sense that in attributing it to a place, one imagines the place as something that need not have it. Part of what it is to *see* the field as bare is to imagine it as otherwise, an aesthetic or imaginative perception as opposed to the ordinary perception of qualities like color, simple givens in the seen environment. Thus,

> S aesthetically or imaginatively perceives x as F if and only if S perceives x and S imagines x to be non-F.

Why does Bhattacharyya insist that my experience of the field as bare, or of the book as missing, is a case of *perceptual* experience, albeit of a special sort? Perhaps because absences in such cases are in a way akin to affordances. Hubert Dreyfus explains that "affordance," a term introduced by J. J. Gibson, is used "to describe objective features of the world in terms of their meaning to the creatures that use them. Thus a hole *affords* hiding to a rabbit but not to an elephant. To us floors afford walking on, apples afford eating, etc." Dreyfus notes that the Gestaltists were "unhappy with the term 'affordance.' They were interested not in our perception of objective features of the world but in how such features are related to the needs and interests of perceivers. So they introduced the term *solicitations*. For Gibson an apple *affords* eating, i.e., is edible, whether anyone is hungry or not, but the Gestaltists add that only when one is hungry does an apple *solicit* eating, i.e., look delicious."[6] The bareness of the field solicits a quest for what is missing from someone who has returned to the field but not from someone there for the first time. It is not that I now see the field simply as an expanse of green grass; rather, it presents itself as a question: "What used to be here?"[7] Here, the phenomenological fact that I am directly acquainted with is the bareness of the field in its greenery. Imagining the place without the attribute is a condition on perceiving it aesthetically or imaginatively as having the attribute, a condition met by regarding the attribute as a solicitation of a particular sort, namely a solicitation to imagine the contrary. The field, insofar as it is seen not merely as a green expanse but as "wearing a bare look," solicits imagining it as adorned.[8] If this is on the right lines then we can understand why Bhattacharyya offers "perceives aesthetically" as an alternative to "perceives imaginatively." For the aesthetic perception of a work of art might well be said to involve a solicitation to imagine the artwork in other, contrary ways and so to see it as more than just an interesting array of colors or colorful things.

TRANSFORMATIONS IN ATTENDING

Bhattacharyya now points out that an experience of imaginative perception can undergo a fundamental cognitive transformation: the perceptual experience transforming itself into a nonperceptual consciousness of absence. To emphasize the importance of relevance in this context, he alters the example:

> To make it more readily intelligible, we may vary the illustration and consider the absence of a beloved person instead of a book that is looked for in a room. When such a person is missed or imaginatively perceived as now absent, there may not be any relevant reference to the locus viz., the room. But one may come to imagine the room as with the person and then realise his absence in reference to this imagined content. To imagine an object in a perceived locus is a special form of imagination in which the present locus is viewed as characterising and not as characterised by the imagined content. The belief in the absence of the object as thus characterised by the locus, the absence here of the imagined room as sentimentally associated with the beloved person, is immediate knowledge but not perception. The absence is not taken to be fact in the present locus; and as the presentness of the absence is not the presentness of any concrete thing, it cannot be said to be perceived. The secondary cognition is conscious nonperception, the room that is perceived by sense being turned into the imagined character of location of the imagined person.[9]

I have in my mind the image of a missing friend, and I imagine her to be in the room, a room itself now imagined and not perceived. This imagining leads me to an immediate but nonperceptual consciousness of the absence of my friend in the room. It is as if I were to close my eyes and imagine her in the room, then open them and, in a flash, understand that the room is devoid of her. I open my eyes and thereby come to know that the room is, as far as my friend is concerned—and that is all I am concerned about—*empty*. This emptiness is not a perceptible attribute of the room, and no deduction is in play. So, it is a case of what Bhattacharyya now calls "conscious nonperception," his term for a new mode of knowing:

> What is known to be now absent is known in the consciousness of not perceiving it, with the belief that it would have been perceived had it been present. Not that it is, therefore, *inferred* to be absent: the consciousness of not perceiving what, it is inferred, would have been perceived is itself no inference and is at once the objective knowledge of the present fact of absence. Present absence by itself is then immediately known in connexion with sense and inference by what may be called conscious nonperception. This has to be recognised as a new mode of knowledge and is comparable with the pure perception of the object, conceived as that to which the object is not given but before which it floats up like an image.[10]

Had my friend been here, I would have perceived her. This counterfactual functions not as the major premise in a syllogism ("I do not perceive her; had she been here, I would have; therefore, she is not here") but as a summary of my epistemic disposition, of my perceptual competence. It is to ascribe to me, we might say, a virtuous epistemic status. My knowledge of my friend as absent consists in the exercise of this epistemic virtue in the present circumstance. I began by seeing my friend *as missing*, where "missingness" is a solicitation. Just as the rabbit sees the hole not as a mere thing but as an escape route, what missingness solicits is the imagined possibility of my friend being present (a possibility that would not have presented itself had I simply stumbled across an empty room without seeking my friend). The perceived solicitation leads me to imagine the contrary situation, and my consciousness is now *of the room* as devoid of the person, rather than *of the person* as missing.

So, the epistemology of absence is two-tiered, involving aesthetic or imaginative perception on one level and a proprietary epistemology in the source of knowledge called "conscious nonperception" on another. Conscious nonperception is that in virtue of which, having seen my friend as missing, I know that the room is empty, and having seen the field as bare, I know that the tree has gone. As a mode of nonperceptual consciousness, conscious nonperception is to be identified with the negative attention referred to in the passage with which we began: attention to an object's absence in a given situation.

Images have an essential role to play in this new epistemic modality, for it is only by way of an imagined contrast that, of all the absences in the room, the one I attend to is the particular and specific absence of my friend. This selection consists in a sort of "pop-out" attention, the object's absence jumping out from the perceived scene and becoming salient by imagining a scene in which it is present and my search is complete. Negative attention as attention to situated absence is a cognitive procedure involving the manipulation of retrieved images and the construction in imagination of contrary scenarios. There is no question of such attention being perceptual attention, for the simple reason that the absent object is not available to the senses and only what is concretely present can be perceptually attended to: "As the presentness of the absence is not the presentness of any concrete thing, it cannot be said to be perceived."[11] It is irreducible to deduction, too; the nonperceptual consciousness is "immediate." The procedure consists rather in a "transformation in attending," other illustrations of which are the spontaneous reversals of the Necker cube and the multistable figures like the duck-rabbit.[12] Such transformations, including our case in which aesthetic or imaginative perception (of the field as bare) is transformed into negative attention (of the tree as gone), are nondeductive restructurings of the field of consciousness.

THE PHENOMENOLOGY OF ABSENCE

You may naturally have been put in mind of Jean-Paul Sartre's later but more famous discussion of a similar case in *Being and Nothingness*.[13] We can better appreciate the originality in Bhattacharyya's idea through a comparison with Sartre. Sartre begins by describing the way in which one's expectation serves to structure the perceptual field into figure and ground:

> I have an appointment with Pierre at four o'clock. I arrive at the café a quarter of an hour late. Pierre is always punctual. Will he have waited for me? I look at the room, the patrons, and I say, "He is not here." Is there an intuition of Pierre's absence, or does negation indeed enter in only with judgement? At first sight it seems absurd to speak here of intuition since to be exact there could not be an intuition of *nothing* and since the absence of Pierre is this nothing. Popular consciousness, however, bears witness to this intuition. Do we not say, for example, "I suddenly saw that he was not there." Is this just a matter of misplacing the negation? Let us look a little closer. It is certain that the café by itself with its patrons, its tables, its booths, its mirrors, its light, its smoky atmosphere, and the sounds of voices, rattling saucers, and footsteps which fill it—the café is a fullness of being. And all the intuitions of detail which I can have are filled by these odours, these sounds, these colours, all phenomena which have a transphenomenal being. Similarly Pierre's actual presence in a place which I do not know is also a plenitude of being. We seem to have found fullness everywhere. *But we must observe that in perception there is always*

the construction of a figure on a ground. No one object, no group of objects is especially designed to be organized as specifically either ground or figure; all depends on the direction of my attention. When I enter this café to search for Pierre, there is formed a synthetic organization of all the objects in the café, on the ground of which Pierre is given as about to appear. This organization of the café as the ground is an original nihilation. Each element of the setting, a person, a table, a chair, attempts to isolate itself, to lift itself upon the ground constituted by the totality of the other objects, only to fall back once more into the undifferentiation of this ground; it melts into the ground. For the ground is that which is seen only in addition, that which is the object of a purely marginal attention. Thus the original nihilation of all the figures which appear and are swallowed up in the total neutrality of a ground is the necessary condition for the appearance of the principal figure, which is here the person of Pierre.[14]

What happens, though, with the realization that the object is absent?

But now Pierre is not here. This does not mean that I discover his absence in some precise spot in the establishment. In fact Pierre is absent from the *whole* café; his absence fixes the café in its evanescence; the café remains *ground*; it persists in offering itself as an undifferentiated totality to my only marginal attention; it slips into the background; it pursues its nihilation. Only *it makes itself ground for a determined figure; it carries the figure everywhere in front of it, presents the figure everywhere to me. This figure which slips constantly between my look and the solid, real objects of the café is precisely a perpetual disappearance;* it is Pierre raising himself as nothingness on the ground of the nihilation of the café. So that what is offered to intuition is a flickering of nothingness; it is the nothingness of the ground, the nihilation of which summons and demands the appearance of the figure, and it is the figure—the nothingness which slips as a *nothing* to the surface of the ground. It serves as a foundation for the judgement "Pierre is not here." It is in fact the intuitive apprehension of a double nihilation. To be sure, Pierre's absence supposes an original relation between me and this café; there is an infinity of people who are without any relation with this café for want of a real expectation which establishes their absence. But, to be exact, I myself expected to see Pierre, and my expectation has caused the absence of Pierre *to happen* as a real event concerning this café. It is an objective fact at present that I have *discovered* this absence, and it presents itself as a synthetic relation between Pierre and the setting in which I am looking for him. Pierre absent haunts this café and is the condition of its self-nihilating organization as ground.[15]

That the individual objects and people in the café dissolve into an amorphous background is clear enough, as is the claim that this is the background for Pierre to appear against as figure. But Pierre is not there, and it is unclear exactly what Sartre is claiming about the consciousness of his absence. Is his absence presented in *intuition* [that is to say, perception], or is it a *discovery*? Pierre's absence "haunts" the café and is not localized in any particular place within the café. Sartre's attention has given the field of consciousness a figure–ground structure but with the unusual feature that the figure is unoccupied. There is an experienced "nihilation" of the café's contents as they dissolve into a background, and there is a "nihilation" of the figure as it is discovered to be vacant.

Here, Sartre is drawing on the work of the Danish psychologist Edgar Rubin. It was Rubin who developed the idea that perceptual experience has a "figure–ground" structure,

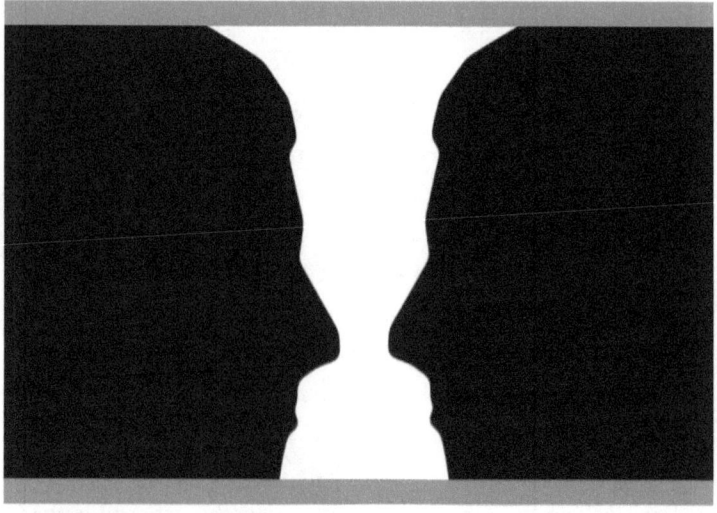

FIGURE 7.1 The Rubin vase: Rubin's illustration of the distinction between figure and ground

the distinction designating the way that part of a perceptual scene "stands out" as a figure against a background.[16] In the illustration of the so-called Rubin vase (figure 7.1), for example, one may see a white vase as figure against a solid black ground or, alternatively, a pair of facing heads against a white background.[17]

The idea is prefigured in William James, though, who wrote, "My experience is what I agree to attend to. Only those items which I notice shape my mind—without selective interest experience is an utter chaos. Interest alone gives it accent and emphasis, light and shade, background and foreground—intelligible perspective in a world."[18] James introduced the idea of a "field of consciousness" with a structure: "In most of our fields of consciousness there is a core of sensation that is very pronounced. . . .

The sensations are the *centre* or *focus*, the thoughts and feelings the *margin*, of your actually present conscious field."[19] Consciousness is a structural field, the structural organization of which is an inherent objective feature of it, not imposed from without by an agentive ego. Sartre rightly observed that, in the case of seeking a friend, interest shapes one's perceptual field of consciousness into a figure–ground structure with the friend as figure and the café as ground and that this is so even if the friend is absent from the café. The object that determines the figure in the field of consciousness does not have to be actually present in the perceived scene.

What is peculiar about the case in hand, then, is that we have an experience in which the figure position is vacant; although the experience is structured to have Pierre as figure and the café as ground, Pierre is not there. As P. Sven Arvidson puts it, the figure is *presented as absent*: "[Sometimes the] theme is evanescent, as in Sartre's famous example of looking for Pierre in the café, and he is not there. Everywhere in the café what is attended to is Pierre presented as absent against the contextual background of the café. It is like searching for the hidden weeds amongst a thick lawn cover. The content in thematic attention is not nothing, it is something. And as soon as the weed presents itself as theme, the previous theme of weed *presented as absent*, an evanescent flickering of thematic content, is replaced by the weed as present."[20] So there must be, as Bhattacharyya emphasizes, two components in the account of consciousness of absence: one that explains the structuring of the field of consciousness into figure and ground and one that accounts for the peculiar phenomenon of the figure being presented as absent.

This is the real significance of the distinction between aesthetic or imaginative perception, on the one hand, and conscious nonperception or negative attention, on the other. We can now

FIGURE 7.2 An example of the Gestalt coherence of a figure (the numeral 4)

understand more clearly why the second component cannot be perceptual. A perceived figure is a unified and consolidated whole, a Gestalt "coherency." That is why it can "stand out" against the background, as in the case of the figure, the numeral 4, in figure 7.2.[21]

Building on the ideas of James and Rubin, Aron Gurwitsch (1901–1973) argues that the topic, which he calls the "theme," is organized as "a unitary whole of varying degrees of richness in detail, which, by virtue of its intrinsic articulation and structure, possesses coherence and consolidation and, thus, detaches itself as an organized and closed unit from the surrounding field."[22] An absence lacks any such structure; it is presented, as Sartre emphasizes, throughout the whole of the café and—or better, according to Bhattacharyya—as lacking in spatial contours or location. Negative attention to absence is "the consciousness of presentness without space-position," and this is why it is a conscious nonperception, why the "cognition of that tree as now absent is no perception, though it is immediate and sense-conditioned cognition of the present."[23]

THE UPSHOT

I began this chapter with a scene of attention. It was a scene in which my attention was captured by the absence of a sought-for object, a missing book. The *absence* of the book, I have argued, is not a Gestalt contexture: the absence is not a figure against a background. Rather, this absence is a *vacancy* in the figure position. The explanation of the idea of being *presented as absent* has required us to go beyond the formal structure of figure and ground, and it is here that we must make an appeal to the imagination, the manipulation of images. Imagining the book as present causes its absence to pop out when juxtaposed against the actually presented perceptual scene. This mechanism of *conscious nonperception* has no clear equivalent in Sartre's phenomenological description of a similar case.[24] Instead, out of Krishnachandra Bhattacharyya's brilliant insights, I have sought to develop a more adequate understanding of the underlying mechanism and thus of the subtle relationships among attention, absence, and imagination.

NOTES

I thank D. Graham Burnett, Justin E. H. Smith, and all the symposiasts at the Princeton History of Science Workshop on the theme of attention, March 19–20, 2020, for their extremely helpful comments and advice.

1. Krishnachandra Bhattacharyya, "Some Aspects of Negation," in *Studies in Philosophy*, vol. 2, ed. Gopinath Bhattacharyya (1914; repr., Calcutta: Progressive, 1958), §6, 207.
2. Bhattacharyya, "Some Aspects," §9, 212.
3. Krishnachandra Bhattacharyya, *The Subject as Freedom* [1930], in *Studies in Philosophy*, vol. 2, ed. Gopinath Bhattacharyya (1930; repr., Calcutta: Progressive, 1958), 1–94; §72, 57–58.
4. On the concept of peripersonal space, see the important new studies in Frédérique de Vignemont, ed., *The World at Our Fingertips: A*

Multidisciplinary Exploration of Peripersonal Space (Oxford: Oxford University Press, 2021).
5. Bhattacharyya, §71, 57.
6. Hubert Dreyfus, "The Myth of the Pervasiveness of the Mental," in *Mind, Reason, and Being-in-the World: The McDowell-Dreyfus Debate*, ed. Joseph K. Schear (London: Routledge, 2013), 15–40, 37n12.
7. We might also think of the idea, popularized by Max Scheler, that certain states of mind are literally visible, that "we certainly believe ourselves to be directly acquainted with another person's joy in his laughter, with his sorrow and pain in his tears, with his shame in his blushing, with his entreaty in his outstretched hands. . . . If anyone tells me that this is not 'perception' . . . I would beg him to . . . address the phenomenological facts." Max Scheler, *The Nature of Sympathy*, trans. Peter Heath (London: Routledge, 1954), 260.
8. This is an early acknowledgment of what has recently been dubbed "mental affordance." See Tom McClelland, "The Mental Affordance Hypothesis," *Mind* 129 (2020): 401–27.
9. Bhattacharyya, §76, 59.
10. Bhattacharyya, §75, 59.
11. Bhattacharyya, §76, 59.
12. P. Sven Arvidson, *The Sphere of Attention: Context and Margin* (Dordrecht: Springer, 2006), 56–85.
13. Jean-Paul Sartre, *Being and Nothingness*, trans. H. E. Barnes (1943; repr., New York: Philosophical Library, 1956).
14. Sartre, *Being*, 9.
15. Sartre, *Being*, 10 (my emphasis).
16. See Edgar Rubin, *Synsoplevede figurer: studier i psykologisk analyse*, vol. 1 (Copenhagen: Gyldendalske Boghandel, Nordisk Forlag, 1915). See also the excellent recent work by Pind: Jorgen Pind, *Edgar Rubin and Psychology in Denmark: Figure and Ground* (Dordrecht: Springer, 2014).
17. See also Lucy Alford, *Forms of Poetic Attention* (New York: Columbia University Press, 2020) for an important application of the figure–ground model in the study of poetic attention. Notice that on Bhattacharyya's analysis, negative attention itself constitutes a form of poetic attention.
18. William James, *The Principles of Psychology* (New York: Henry Holt, 1890), 380–81.

19. William James, *Talks to Teachers on Psychology and to Students on Some of Life's Ideals* (1899; repr., Cambridge, MA: Harvard University Press, 1983), 18.
20. Arvidson, *The Sphere of Attention*, 72. Again, in an earlier essay, Arvidson writes, "Is the pill bottle in the next room? Where is it? Now the pill bottle is thematic. . . . Attention is much more complicated than my description suggests. For example, in the last part the pill bottle is thematically *presented as absent*." See P. Sven Arvidson, "Experimental Evidence for Three Dimensions of Attention," in *Gurwitsch's Relevancy for Cognitive Science*, ed. Lester Embree (Dordrecht: Springer, 2004), 155.
21. Reproduced in Aron Gurwitsch, "The Field of Consciousness," in *The Collected Works of Aron Gurwitsch*, vol. 3, ed. Richard Zaner and Lester Embree (1964; repr., Dordrecht: Springer, 2010), 130.
22. Gurwitsch, "The Field of Consciousness," 139.
23. Bhattacharyya, §79, 60–1, §75, 59.
24. In an earlier work, Sartre describes the imagined Pierre as "irreal," such that "this Pierre in relation to whom the situation is defined is precisely Pierre *absent*." See Jean-Paul Sartre, *The Imaginary: A Phenomenological Psychology of the Imagination*, trans. Jonathan Webber (1940; repr., London: Routledge, 2004), 187.

8

DISPATCH FROM THE JHĀNA WARS

Attention Practice in Online Buddhism

JOHN TRESCH

A SCENE OF ATTENTION: REMOTE CONGREGATION

At the end of August 2020, as the United Kingdom was sputtering open from its first COVID-19 lockdown, I took a train north from King's Cross to Carlisle. Despite the sunshine, the trip was eerie; it was my first travel in six months. Worried about sharing an enclosed space, passengers sat distanced from one another. I continued west on a commuter train to Annan, Scotland. I stocked up at the Aldi and took a taxi three miles onward to a tiny town called Brydekirk, where I'd booked a "bothy," a long, low cabin above a garage on a hill overlooking the river Annan. It could have been anywhere. I just needed a quiet place with decent Wi-Fi. I was there for a twelve-day silent jhāna meditation retreat starting that night. I was alone and would be joining the others by Zoom.

Jhāna is a form of Buddhist meditation that depends on heightened, sustained attention. In the last decade, jhāna has been increasingly practiced and discussed among North American and European "convert" Buddhists, dispersed communities that are largely White, educated, and middle class.[1] Though part of the wider scene of Western Buddhism, jhāna differs from the better-known practices of "mindfulness" and "Vipassana" meditation in its nearly exclusive focus on concentration (*samadhi*) and the attention it requires. The teaching of both mindfulness and Vipassana emphasizes acceptance of whatever "naturally" arises in the mind and body. In jhāna, the attention is more sustained and more directed, and the states it induces arise because of the meditator's effort and intention.

According to the leader of the 2020 retreat I attended, Leigh Brasington, jhānas are "a very useful way of preparing your mind, so you can more effectively examine reality and discover the deeper truths that lead to liberation."[2] Like other contemporary jhāna teachers, Brasington has participated in discussions and collaborations with neuroscientists and philosophers in which attention practices are seen as necessary for understanding the mind.[3] In his book *The Attention Revolution*, Alan Wallace, an ordained Tibetan monk, observes, "While scientists have tried to understand the mind by means of objective, third-person inquiry, contemplatives for millennia have explored the mind by means of subjective, first-person inquiry. Such investigation into the nature of the mind is meditation, and truly effective meditation is impossible without focused attention."[4] Wallace, also the founder of the Santa Barbara Institute for Consciousness Studies, is the meditation teacher and codirector, with brain researcher Cliff Saron, of the Shamatha Project, a months-long residential retreat that studied participants' minds and brains before, during, and after intensive attentional training.[5]

Evan Thompson, a philosopher who has been deeply involved in both organizing and writing about the interactions between Buddhism and contemporary science, has recently reflected on what he sees as certain pitfalls of "mindfulness mania" and "neural Buddhism." Thompson urges researchers to be more specific about the practices that they study as "meditation"; this demands an in-depth knowledge of the traditions of which they are a part. Further, he argues, "we need to move from focusing just on the brain to examining how cultural practices orchestrate the cognitive skills that belong to meditation."[6] My chapter heeds Thompson's suggestions. I introduce the jhānas' particular forms of attention and the knowledge of the mind they imply through a historically grounded first-person account of their current cultural settings. I put particular focus on the controversies that have been sparked by the revival of jhāna practices, referred to as the "jhāna wars" by one commentator.[7] Further, as the return of the jhānas has taken place in the same years as an intensification of digital communication, I consider how their practice and the debates around them highlight more general connections between reinvented traditional forms of mental and spiritual cultivation and lives increasingly lived online.

DROPPING IN

After unpacking and setting up the tiny kitchen space with canned soup, fruit, ramen noodles, and instant coffee, I took a stroll through the village. There were no shops, and the one pub had recently closed down; the streets, like the nearly windowless church at the top of the main street, were silent, almost deserted. In my garage cabin, I set up my laptop on a low table like a dais in front of the pillows and blankets where I'd sit. The setting sun lit up the trees along the river through the small window behind my screen. After some confusion I found the BT Wi-Fi, logged on, and turned on my out-of-office message. As night fell, I stretched out on the bed to nap and wait for the course to begin.

The retreat was based in California and Florida. At midnight, I clicked the Zoom link. The bright blue chessboard blinked to life, showing each of us in our squares with varying configurations of blankets, bookshelves, Buddhist icons, partners, dogs, and cats in the background. The teachers, Leigh Brasington and Rachael O'Brien, gave brief welcomes. There were about twenty participants, all with some retreat experience, in a two-to-one ratio of men to women. We ranged in age from the late twenties to the midsixties, with the largest group being those in their midthirties. Most were White, a couple were Asian, and we were scattered across the United States and Europe, except for one person joining from Australia at some horrendous hour.

After welcoming us, Rachael—a brisk, affable lawyer in her early forties—recited the traditional opening chant. In our makeshift cells, we repeated her words together: "I take refuge in the Buddha, the *dhamma*, the *sangha*" (the enlightened one, his teachings, and the community of practitioners).

We then closed our eyes and began our "grounding practices," in whatever form we were most comfortable with—whether concentrating on the breath under the nostrils or at the abdomen, attending to bodily sensations, or cultivating the heart-feeling of *metta* (benevolence or loving-kindness). Any of these, Rachael said, was good for concentrating the mind for jhāna. She spoke occasionally, reminding us to come back to the object of meditation. "If your mind wanders away, recognize it, release it, and relax. Now smile, and return to the object of meditation." Through my laptop's tinny speakers, the sound of her voice and breath reminded me that my own effort to focus and quiet down was echoed by my fellow retreatants. This strange dark chamber was linked to all these other distant rooms. After an hour, she rang a chime. Heads bowed to screens. *Leave meeting*, I clicked.

For each of the next twelve days, we gathered in our sweaters, flannels, robes, and pajamas for two group sittings and a dhamma talk—lectures on practice and what some call the "cosmological" theory of dependent origination, alternately led by Leigh and Rachael. Over the course of the retreat, each student also had two thirty-minute sessions alone with each teacher. That was the only time I spoke. Between sittings, I took walks through and beyond Brydekirk, nodding to the dog-walkers I occasionally passed on the wooded path by the river, the three little girls on bikes chasing one other on a repetitive circuit through the town, a farmer in a tractor overtaking me on a road.

By the second day, I fell into a pattern of waking at six or seven, sitting for an hour, and then continuing until 9 or 10 p.m., with breaks for breakfast, walks, lunch, and dhamma talks. By the fifth day, I found myself back in jhāna territory. Even over wavering Wi-Fi emanations, the teachers and sitters cohered as a group, supporting one another's practice.

My glance at this online retreat and other contemporary settings for the practice of jhāna is part of a historical and ethnographic study of interactions between Western Buddhism and contemporary technoscience. I'm writing as a participant observer. My research started in 2005 when I attended a meeting of Mind & Life, an organization founded by the Chilean biologist and philosopher Francisco Varela. Bringing together meditation teachers, monks, and scientists, Mind & Life has become an institutional base for the neuroscience of meditation.

I've spent time at various conferences and retreats, including several weeks in 2007 in the Rockies with the Shamatha Project, where a mobile laboratory examined the physiological, cognitive, affective, and behavioral effects of concentration practice, taught by Alan Wallace, during a three-month residential retreat; Cliff

Saron's team at the University of California, Davis, is still analyzing their terabytes of data.[8] Over New Year's in 2015, I joined a retreat conference on "contemplative development" at the Barre Center for Buddhist Studies in Massachusetts, organized by Willoughby Britton and Jared Lindahl from Brown University. One of the participants was Leigh Brasington.[9]

At the time, I knew little about the jhānas. I'd heard the word, but since 1999 I had been daily practicing Vipassana meditation as taught by S. N. Goenka. Though Goenka mentions that jhāna can be practiced on long retreats, the Goenka school doesn't foreground them and strongly discourages experimenting with other traditions. But for some time I'd felt uncertain about my practice. I'd been having bad headaches on retreats, and my fieldwork had exposed me to a wide range of contemporary Buddhist teachers and techniques.

At the Barre Center, where participants meditated and discussed their research, along with the setbacks, dead ends, and crises they experienced in their meditation, I met with Brasington. He's a tall, bearded man of about sixty who used to design databases and often lives in Northern California; he delivers stories and instructions in a light Southern accent in a self-effacing, funny, appealingly folksy manner. I told him how my headaches had been increasing along with my concentration and that I hoped that pushing through with Vipassana—involving an awareness of breath (*anapana*) and a systematic body scan—would eventually bring some kind of breakthrough. If I concentrated harder, the headaches would end, and the periods of quiet, peace, and revealing reflection I'd savored on many retreats would seep more thoroughly into daily life.

Brasington told me I had it backward. "Unless you're already relaxed, you can't concentrate." He suggested I take a course with him.

I've now done six jhāna retreats, including four with Brasington and his co-teachers. Although I'm in kindergarten with the jhānas, my limited experience already reveals some patterns and fault lines. The philosophical investigation of attention—its varieties, components, and relationships to judgment and knowledge—is at least as developed in Buddhist traditions as it is in Western philosophy and psychology. A key plank of the "eight-fold noble path" is *sama samadhi* (right concentration): knowing what to attend to and how. Yet beyond theoretical definitions or phenomenological investigations of attention, for many Buddhists, especially Western converts, knowledge of attention also involves a range of practices for developing attention. Debates about their meaning now play out not in monastic settings but in online forums and on meditation cushions in practitioners' living rooms.

FORM AND FORMLESS

The Pali term *jhāna* (*dhyāna* in Sanskrit) simply means "meditation," though it has an etymological link to *jhāyati* ("fire," or "conflagration"); in Buddhist suttas, when the Buddha tells his followers to "go meditate," he says, "Go practice jhāna." *Jhāna* now usually appears in the plural: "the jhānas" refers to a series of eight states of deep concentration, each more refined than the last. The jhānas are said to offer rest, relaxation, and intense, nonsensuous pleasure. They are also said to prepare the mind for the practice of insight, or *Vipassana*: a clear view into the nature of reality that leads to enlightenment or liberation from delusion and suffering.

The jhānas are often described as distinct "realms," places your mind can go. The first four are presented sequentially in the Buddha's Pali discourses.[10] He described the first jhāna thus: "Quite withdrawn from sensuality, withdrawn from unskillful mental qualities, [the meditator] enters and remains in the first jhāna: rapture and pleasure born from withdrawal, accompanied by directed thought and evaluation. He permeates and pervades, suffuses and fills this very body with the rapture and pleasure born from withdrawal."

The phrase "directed thought and evaluation" (*vitakka* and *vicara*) is sometimes translated as "directed and sustained attention." Either translation implies a significant degree of focused concentration. But I've also heard the phrase translated merely as "this and that," meaning "everyday speech" or "ordinary internal dialogue"; this reading implies a less acute state of concentration.

The most important signs of the first jhāna are *piti* and *sukka* ("rapture" and "joy," "happiness," or "sweetness"). In Brasington's teaching, *piti* is any pleasurable sensation, often taking the form of a tingling or energetic tremor, like a current of electricity, while *sukka* is sweetness and happiness, even a voluptuous enjoyment. In the first jhāna, the body is absorbed in these sensations, just as in the sutta's analogy, the powder kneaded into soap by a bath man is suffused with moisture. Like the other jhānas, the first is presented as a direct reward of meditation, a "fruit of the contemplative life, visible here and now."

The next three jhānas take a similar form. In the second, rapture and sweetness remain but with a greater "unification of awareness, free from *vitakka* and *vicara*" (with either a less "directed" attention or a silencing of internal dialogue), accompanied by "internal assurance." *Piti* recedes into the background and *sukka* becomes more prominent. The third jhāna sees "rapture" (*piti*) drop away, while "joy" (*sukka*) remains: "With the fading of rapture, [the meditator] remains equanimous, mindful, and alert, and senses pleasure with the body." In the fourth jhāna, *sukka* also drops away, revealing the "purity of equanimity and mindfulness": "Just as if a man were sitting covered from head to foot with a white cloth so that there would be no part of his body to which the white cloth did not extend; even so, the monk sits, permeating the body with a pure, bright awareness." Achieving the pronounced calm of the fourth jhāna is an important stage for conducting the introspective questioning that leads to insight. This may involve inquiry into the "three characteristics" or "marks of existence" of any phenomenon: its changing nature (*anicca*), its lack of any fixed essence or soul (*anatta*), and its unsatisfactoriness (*dukkha*). Thorough realization of any of these marks is a "door" to enlightenment.

The first four jhānas are called the *rupa* ("form") jhānas since they involve perception of the physical body. The next four, also mentioned in the suttas, though only later named and numbered, are called the *arupa* ("formless") jhānas, "peaceful immaterial liberations transcending material form."[11] The fifth is defined as "limitless space," the sixth as "limitless consciousness," the seventh as "nothingness" or "void," and the eighth as "neither perception nor nonperception."

These "formless" realms are famously difficult to describe, even by those who have definitely encountered them (not me). They are increasingly subtle, requiring an ever calmer and more stable mind. Some teachers allow that in these states of "absorption," an awareness of the physical body and senses remains; for others, their full realization involves complete "seclusion" from bodily sensation, the ceasing of awareness of the external senses. Another fruit of the jhānas is said (though not by Leigh Brasington) to be *abhijna* ("psychic powers" involving remembering past lives, telepathy, and passing through solid substances). One pair of jhāna teachers semi-jokingly describes jhāna practice as

"Jedi warrior mind training." A participant of my 2020 retreat wore a different Star Wars T-shirt most days.

Even if the sequence of jhānas leads to states that "transcend material form," the concrete places where they're taught, the actual rooms in which students and teachers meet, also deserve consideration. My first jhāna retreat, in 2015, was on a retrofitted farm in Indiana. Before that, I had attended Vipassana and insight retreats in groups of between fifteen and 200 people at dedicated retreat centers but also at a children's summer camp in France, a college gym, a nuns' retirement home, a Theosophists' estate outside Amsterdam, and even a funeral parlor in Chicago, purified of harmful forces by a looped recording of chanting played overnight.

Both Vipassana and jhāna retreats often require at least ten days of silence, without phone, internet, books, or talking except with teachers. They start with the vows of "taking refuge," and sessions open and close with the sound of chimes, a bell, or a gong. These ritual time markers and the physical presence in a single room of the group, teacher, and volunteer staff provide powerful encouragement and soft surveillance. During days of silent sitting for seven to twelve hours, knowing that others are hard at it on all sides helps you stay focused; knowing that someone might be watching if you open your eyes or sneak out for a break is a spur to diligence.

As in other kinds of mental training in diverse traditions worldwide, individual attention is a social achievement. Beyond these disciplinary mechanisms, teachers comment in supernatural or quasi-supernatural terms on the quality of silence that comes from sustained group practice, from "tuning up the atmosphere." A jhāna retreat seems to create special conditions that allow people to drop into jhānas one by one and together.

GLIMPSES OF JHĀNA

On my first jhāna retreat, I began by practicing *anapana*, attention on the breath going in and out of my nose, with a relaxed, gentle focus. In a one-to-one meeting, one of the co-teachers, Jay Michaelson, told me that "*anapana* is really dumb. Don't make your attention too sharp. Just hang out there and relax." This was a change from my earlier habit of hypervigilance—which had led to headaches. He also told me it was OK to medicate if they returned: "You can be an Advil yogi," he joked.

As usual on retreat, for the first day or two the restlessness of my mind was annoyingly vivid—the constant narration, evaluation, speculation, longing, and complaint described as "monkey mind." Around the third or fourth day, I had sessions during which the "five hindrances" (doubt, restlessness, sleepiness, craving, and aversion) lifted. I was left with a pleasantly neutral, expansive feeling of openness and calm. Drawing on something I'd overhead another meditator say before the retreat began, I thought of this as "access consciousness." I had been there on previous retreats but didn't have a name for it, nor had I seen it as the doorway to jhāna.

Following Brasington's instructions, starting on the fourth and fifth days I began experiencing something resembling the jhānas, as I have described them. A pleasurable tingling sensation (*piti*) arises; you turn your attention to the "pleasantness of the pleasant," possibly stoking it with a feeling of benevolence (*metta*) until the sensation spreads and engulfs the whole body, accompanied by an exhilarating feeling of joy . . . and so on, through the first two or three jhānas, more tentatively into the fourth. In the last couple of days of the retreat I made a few faltering attempts to get a feel for the fifth and sixth.

Though I'd had similar experiences on Vipassana retreats, it was a change to have them defined as regularly recurring states, lined up on a progressive scale. It was also a change to seek them out and try to linger in them or to move at will from one to the next. My previous teachers insisted on simply observing with equanimity, giving "no importance" to any experiences, not seeking to prolong or get rid of them. Now I was being told to sustain and cultivate particular states and, at least in the first three jhānas, to differentiate them, luxuriate in them.

One afternoon late in the retreat, after remaining for some time with something that felt like the "pure, bright awareness" described as the fourth jhāna, I stood up and walked out of the rehabbed farmhouse into the orchard and the afternoon sun. I found my mind very calm; my internal discourse not merely slower and quieter but nearly silent; the colors, lights, and forms of the Indiana farmland lit up with a gorgeous, serene intensity. A resonant, quietly wondrous feeling lingered into the next day. I remain unsure whether I actually had been in the fourth jhāna, and I can't put a name to what insight I derived from it. But the experience was worth the effort.

On another jhāna retreat the following year with a different teacher, I was encouraged to strive for a greater state of concentration and await a specific sign (*nimitta*), a glowing disk in the visual field, before entering the first jhāna. The *nimitta* didn't show up, and I got a terrible headache. I've heard another teacher insist that the question of whether you've "really got it" is a distraction from the real aims of jhāna practice: keeping the mind agile and playful by increasing and decreasing the degree of intensity, absorption, and enjoyment.

In my 2020 online retreat, Brasington's co-teacher, Rachael O'Brien, helped with the teaching and interviews. She has practiced extensively with an American Theravadin monk, Bhante

Vimalaramsi, but shared Brasington's admiration for his primary teacher, Ayya Khema (1923–1997). Born in Berlin and subsequently dislocated by the Nazis, Ayya Khema later became ordained as a Theravadin nun and taught mostly in California, Australia, and Germany. She was evoked throughout the course, as an exemplary, empathic teacher for whom the jhānas were the core practice in a life committed to ending the suffering of herself and others.

The retreat had originally been planned for a Franciscan monastery in Florida near Rachael's home, but COVID-19 forced it online. Previous retreats I had taken offered silent isolation, often in a remote and beautiful or unusual place, along with a continuously present, dedicated community. Though the latter was missing, the greater independence of the online setup let me tailor my schedule. I sat in the mornings and afternoons and took long walks on breaks. I ate my meals outside on the small patio, reading a book subtitled *Meditations on Emptiness and Dependent Arising*.[12] Reading was a continuation of practice, as was attending to my movements, the rustling trees, the rushing current of the river, and the cool breeze on my walks. The one-to-one conversations I had with Rachael and Leigh felt a bit rushed and disjointed, as interactions on Zoom often do, but in group sittings, with our cameras on, there was a firmness comparable to an in-person retreat.

If the internet was my means of accessing the group, it could also be a way to disengage; it made it possible to join in the collective practices of attentional development but also posed a threat to this hard-won concentration. I didn't have the option of switching off entirely, and every time I turned on my computer for group sitting I felt tempted to check my email or visit news sites and blogs. At lunch one day, I peeked at my inbox and saw a curious subject heading in an email from an old friend.

I read the email and clicked the link to an article my friend had included. One of my teachers had died suddenly—an anthropologist, activist, and theorist, someone I looked up to. I felt viscerally shaken; my attention scattered. A new undertone of gloom, fear, and doubt appeared.

Discomposed, I took a walk. I regretted piercing the retreat's bubble. But I reminded myself that this was a refuge, a place to face a such a fact, however incomprehensible. It was proof of impermanence, the fleeting nature of achievement, of life. *Anicca, anatta, dukkha.*

When I logged on later for the group sitting, I felt sad and tender toward the teachers and students, each in our fragile boxes navigating emptiness. Distraction was a click away. Stillness was also at hand.

ELECTRONIC AFFINITIES

Much like the "elective affinities" Weber saw between Protestantism and capitalism, the recent spread of individualized, consumer-friendly, easily accessed meditation resonates with the rise of digital media. Beyond the uptake of mindfulness in Silicon Valley, the internet has made meditation much more widely available. Some critics link the recent wave to neoliberal aims: "McMindfulness" places the responsibility of dealing with anxiety, injustice, or deprivation onto the shoulders of the individual, and "accepting reality as it is" may be another name for passivity and quietism.[13] Recent recontextualizations of Asian traditions follow longer tendencies that David McMahan has summarized as "Buddhist modernism."[14] Since the nineteenth century, Western reframings of Buddhism have often downplayed ritual and metaphysical beliefs such as karma, deities, and rebirth, concentrating instead on individual psychology. They emphasize aesthetic responses to nature, in line with Romanticism. They also present Buddhism's empiricism, causality, and rational explanation as compatible with scientific naturalism.[15]

But in the last two decades, the rapid availability of information about diverse traditions, including videos of dharma talks and guided meditations and frank discussions of retreat experiences, has introduced new tendencies to Western Buddhism. A new generation of teachers has emerged, many of them Gen X or millennials, who are largely independent of established institutions, rules, and structured initiations. They take a mix-and-match "consumer approach" toward teachings; they meet not in locally bound communities but on discussion boards and podcasts. Naming this turn "Buddhist postmodernism," the anthropologist Ann Gleig notes a "shift from a counterculture hippy mentality to an urban technologically savvy subculture mindset." Meditation is often presented as "tech" or "brain hacking."[16] Scientific studies of meditation are seen as aids to understanding practice; according to Jay Michaelson (who assisted Leigh Brasington in my first jhāna course), "the science changes how the dharma is even to be understood."[17] On retreat, he and Brasington mentioned neuroscientific findings and evolutionary theories about what the brain might be doing during the jhānas.

A landmark in this new wave was Daniel Ingram's 2007 open-source publication, *Mastering the Core Teachings of the Buddha: An Unusually Hardcore Dharma Book*.[18] The book presents itself as a no-nonsense antidote to the refusal of most Western Buddhist teachers to discuss enlightenment and the stages leading to it, whether from modesty or ignorance. Ingram's Dharma Overground website is an anchor for "pragmatic dharma," where posters compare experiences and discuss "what works" without deference to traditional authorities. Ingram has also since

2020 been developing the Emergent Phenomenology Research Consortium, "a multidisciplinary, multinational alliance of researchers, clinicians, and patrons who share a vision of bringing scientific methods and clinical sensibilities to the rigorous, ethical, ontologically agnostic study of emergent phenomena." *Emergent phenomena* is the group's term for "effects and experiences, both potentially beneficial and challenging, that would often be referred to as 'spiritual,' 'mystical,' 'energetic,' 'magical,' etc. in less scientific and less clinical contexts."[19] Another reference point for very online Western Buddhists is the Buddhist Geeks website, started in 2006, which features interviews with scientists, meditators, and philosophers.[20] The site has now reframed itself as a "metasangha": "the culmination of a decade of working on figuring out how to bring Sangha from analog to digital space."

These "disruptors" of "consensus Buddhism" often employ the language, tools, and strategies of Silicon Valley. A current of entrepreneurship is visible in meditation apps such as Headspace and Buddhify and among meditators who offer one-on-one meditation consultations and training for a fee.[21]

Members of this new generation have critiqued what they see as the misjudgments of their forebears: their vagueness or secrecy about occasional negative effects of meditation, their dogmatic adherence to teachers or schools, and, in some cases, their indifference to the Buddha's actual words. Despite the iconoclastic stances, many of the new generation refer (in a twist on Luther's *sola scriptura*) to early Buddhist suttas. There is thus a combination of innovation, traditionalism, and even fundamentalism in their approach.

The jhānas fit snugly into this new configuration. If meditation is a technology, jhānas are a significant upgrade, available only to a limited group of discerning, well-informed users. Like

Ingram, Brasington has lent his brain to fMRI and EEG studies. He and other teachers quote chapter and verse to explain and defend the jhānas, such as the Buddha's words in the *Majjhima Nikaya*: "I say of this kind of pleasure that it should be pursued, that it should be developed, that it should be cultivated, and that it should not be feared."[22] By selectively retrieving Buddhist scripture and pursuing its esoteric reaches, jhāna practice appears both less dogmatic and more authentic than earlier phases of Buddhist modernism.

We might speculate about further affinities between the jhānas and online culture. The topology of the jhānas resembles the architecture of the internet; in both cases, "geeking out" means "getting into the zone," moving in a trance-like state from one self-contained domain to another. The jhāna series also resembles the narratives of superhero multiverses and video games. The transition from one jhāna to the next involves "leveling up," while losing concentration sends you back to level one: *game over*. Whether online or on retreat, the (brain) hacker undertakes the hero's journey without leaving the basement—or garage.

For what it's worth, I encountered the jhānas when I needed them. After breakthroughs in my first years of practice, I felt stalled. The jhānas pushed me off a plateau, giving me a welcome sense of getting somewhere: open fields for further practice.

LINES IN THE SAND

The jhāna wars play out against the generational tensions I have described here. A first major point of contention is whether the jhānas should be practiced at all. As "constructed" states attained through will and effort, the jhānas run counter to a basic orientation of much contemporary Western Buddhism. Mindfulness, insight, and Vipassana forms of meditation all insist on an attitude of neutrality or passivity toward experience, observing without reaction whatever presents itself. Nothing should be added or sought beyond the present moment. In these communities, there is also a taboo against discussing attainments—unusual states, breakthroughs, landmarks—and even progress. Such experiences might be discussed with a teacher, but sharing them with other students could lead to ego inflation, envy, competition, or disappointment. The discussion and comparison of personal experiences in jhāna scenes, especially online, break this code.

Another reason to avoid the jhānas concerns enjoyment and pleasure. Meditation can be very difficult; pain in the joints, boredom, sleeplessness, and fatigue are unavoidable on long retreats. The jhānas offer relief, refreshment, even "bliss." Some Vipassana teachers offer strict warnings against becoming attached to pleasurable sensations; for them, the rapture of jhāna may be a dangerous and addictive trap, to be presented with enormous caution. Reciprocally, some self-proclaimed "jhāna junkies" disparage Vipassana without jhāna as "dry insight"—as a masochistic denial of harmless enjoyment.

For these reasons, among many insight teachers, the jhānas are kept quiet, saved for private teaching on long retreats, or avoided entirely. I have friends in the Goenka school who are intrigued by the jhānas but remain reluctant to seek them out. I know other meditators who have heard about the jhānas through the currents discussed in this chapter but find the online scenes so alienating—with their emphasis on "technique" and "attainments" and their saturation with techie males with Darwinian priors—that they keep clear.

If one does decide to practice the jhānas, other disagreements appear regarding interpretation and practice. For instance, what exactly is "directed thought and application?" How is "rapture" different from "sweetness"? Do the senses shut down entirely in the *arupa* jhānas?

The most fundamental interpretive divide lies between teachers who see the jhānas as nearly impossible to obtain by anyone but extremely dedicated monks and nuns—the advocates of "heavy jhāna"—and those who see them as relatively easy for anyone to access—sometimes called, in Brasington's term, "Jhāna Lite." Many of the former find justification for their stance in the long tradition of commentary and monastic practice in the centuries after the Buddha lived. In particular, they accept the authority of the *Path of Purification* (*Visuddhimagga*) by the fifth-century Sri Lankan monk Buddhaghosa. This systematic treatise argues that absorbed meditation is not essential for liberating insight. It describes the jhānas but posits the *nimitta* as a crucial threshold for entering the first jhāna—a rare and difficult attainment achievable by only one in a million meditators.[23]

The heavy jhānas (*Visuddhimagga* jhānas) require weeks or months to access and explore. The Burmese teacher Pa-Auk Sayadaw trains students this way, as does the London-born monk, Ajahn Brahm, who argues that deep concentration, in which experience of the body and all senses but the breath disappear, is necessary for liberation. He denounces "teachers today [who] present a level of meditation and call it *jhāna* when it is clearly less than the real thing." Shaila Catherine, a meditation teacher based in Silicon Valley, also teaches these heavy jhānas, though she allows for the validity of Jhāna Lite.[24]

Brasington unapologetically teaches the lighter, more accessible form of jhāna. As support, he frequently cites the words of the Buddha in the Pali suttas; among the most important evidence he offers for his interpretation is the repeated scriptural reference to ordinary people who master the jhānas and attain liberation. If the jhānas were so easily accessible in the Buddha's time, how could they have become more difficult now? Like other teachers, he sees the *Visuddhimagga* as making unjustified additions to the original scripture. According to Bhante Henepola Gunaratana, what the suttas say about jhāna "is not the same as what the *Visuddhimagga* says." A teacher in San Diego from the Thai Forest tradition, Thanissaro Bhikku (also known as Ajahn Geoff), derides the fixed attention demanded by the *Visuddhimagga* as "wrong concentration"; he says that it makes meditators "psychologically adept at dissociation and denial." Rachael O'Brien's primary teacher, Bhante Vimalaramsi, calls the *Visuddhimaggha* jhānas "hypnosis" and sees their promise of superpowers as "really dangerous." For him, jhāna is a light state of collected attention, available even in daily life.

STRIVING WITHOUT GOAL

Three major positions thus define the jhāna wars: no jhānas, heavy (*Visuddhimagga*) jhānas, and sutta jhānas (Jhāna Lite). In the background of these disagreements stand differing views of the relationship between Buddhist attentional practices and wider cultures of modernity. Opponents of the jhānas present meditation as either a simple, fuss-free method (mindfulness or bare awareness) or as a rigorous ascetic path; jhāna appears as an unnecessary complication or as a frivolous departure from the hard, painful work of insight. The defenders of the *Visuddhimagga* see the jhānas as the fruit of prolonged, undeviating practice appropriate and accessible only to those who renounce worldly affairs. In contrast, advocates of the sutta jhānas, such as Brasington and his co-teachers, see these states of heightened attention as essential to the path of liberation and readily available: a goal within reach for laypeople.

The most rigorous and disciplined path may not be the most direct. Echoing a paradox heard in many religious traditions, Brasington has told students that the only way to enter the jhānas is by not wanting to enter the jhānas. Walking the fine line of striving without attachment seems particularly difficult if meditation is presented as "tech" or "brain hacking." Framed as a means of reaching well-defined, individually satisfying ends (such as pleasure, reduced stress, or increased productivity), jhāna practice, like other current forms of meditation, may well reinforce the attitudes and aims of late capitalism: to become resilient, to collect interesting experiences, to control, to disrupt.

Yet if digitally damaged attention is a central problem of our time, the jhānas may be a solution: a way of reclaiming and learning to redirect the mind. They offer attention as a site for discovery and cultivation—a first-person experience of self-directed plasticity. The collective development of a sustained, supple, and directed attention is a promising counterdiscipline against the addictive encroachments of the attention economy, even with the jhānas' current enmeshment with digital culture.

On the last day of my 2020 retreat—it was late afternoon in Brydekirk and morning in California—we met once more on Zoom. Leigh warned us about returning to our ordinary lives: "If it seems a little crazy out there . . . that's because it *is* crazy out there!" Don't operate any heavy machinery, he said; don't try to clear your whole inbox right away. If you want to keep your attention strong enough to continue to access the jhānas, you'll need two hours of sitting per day, but even half an hour daily should keep you in shape for your next retreat. Rachael also advised reducing ambient distractions: don't put the radio on in the background, lay off the internet.

She led us in a final *metta* meditation. As we closed our eyes, she told us to cultivate a sense of appreciation and gratitude for the opportunity to practice together. We filled ourselves with a wish for all beings to be safe, happy, and free from suffering, starting by putting attention on our ourselves and gradually widening the focus to include to the whole world, transmitting the wish outward. We smiled and waved at the screen; there was a flicker and then a neutral square as the link was broken. *The meeting has been ended by the host.*

NOTES

The author thanks Graham Burnett, Justin Smith, Leigh Brasington, Rachael O'Brien, Jay Michaelson, Clare Carlisle, Daniel Stuart, Elisabeth Hsu, Matthew Drage, Natasha Schüll, Caleb Smith, and all the participants in the workshop on attention.

1. Natalie Quli, "Multiple Buddhist Modernisms: Jhāna in Convert Theravāda," *Pacific World* 10, no. 1 (2008): 225–49; Erik Braun, "The United States of Jhāna," in *Buddhism Beyond Borders: New Perspectives on Buddhism in the United States*, ed. Scott Mitchell and Natalie Quli (Albany: SUNY Press, 2015), 163–80.
2. Leigh Brasington, *Right Concentration: A Practical Guide to the Jhānas* (Boulder, CO: Shambhala, 2015), 6.
3. Michael R. Hagerty et al., "Case Study of Ecstatic Meditation: fMRI and EEG Evidence of Self-Stimulating a Reward System," *Neural Plasticity* (2013), https://doi.org/10.1155/2013/653572; on attention as the object of study, see Antoine Lutz et al., "Attention Regulation and Monitoring in Meditation," *Trends in Cognitive Sciences* 12, no. 4 (2008): 163–69.
4. B. Alan Wallace, *The Attention Revolution: Unlocking the Power of the Focused Mind* (New York: Simon & Schuster, 2006), xi.
5. For an overview of the project's aims, methods, and findings to date, see "The Shamatha Project," Saron Lab, accessed August 31, 2022, https://saronlab.ucdavis.edu/shamatha-project.html.
6. Evan Thompson, *Why I Am Not a Buddhist* (New Haven, CT: Yale University Press, 2020), 159.
7. Justin Merritt, "Jhana Wars! Pt. 1. What Is Jhana Really?," *Simple Suttas* (blog), May 9, 2013, https://simplesuttas.wordpress.com/2013/05/09/jhana-wars-pt-1-what-the-heck-is-jhana-a-first-pass/.

8. See John Tresch, "Experimental Ethics and the Meditating Brain," in *Neurocultures*, ed. Francisco Ortega and Fernando Vidal (Frankfurt: Peter Lang, 2011), 49–68; "Ecologies of 'Mind,'" (Buddhism, Mind, and Cognitive Science conference, University of California, Berkeley, March 2014).
9. Leigh Brasington, *Right Concentration: A Practical Guide to the Jhānas* (Boulder, CO: Shambhala, 2015); Jared R. Lindahl et al., "The Varieties of Contemplative Experience: A Mixed-Methods Study of Meditation-Related Challenges in Western Buddhists," *PloS One* 12, no. 5 (2017): e0176239.
10. *Samaññaphala Sutta: The Fruits of the Contemplative Life*, DN 2 PTS D i 47, trans. Thanissaro Bhikkhu, Access to Insight, 1997, https://www.accesstoinsight.org/tipitaka/dn/dn.02.0.than.html.
11. "Numbered Discourse 9.41: 4. The Great Chapter: With the Householder Tapussa," Sutta Central, https://suttacentral.net/an9.41/en/sujato.
12. Rob Burbea, *Seeing That Frees: Meditations on Emptiness and Dependent Arising* (Devon: Hermes Amāra, 2014); Rob Burbea, "Retreat Dharma Talks: Practising the Jhānas," *Dharma Seed*, December 2019, https://dharmaseed.org/retreats/4496/ (recordings of dharma talks from a twenty-three-day jhāna retreat at Gaia House in Devon, UK).
13. Ronald E. Purser, *McMindfulness: How Mindfulness Became the New Capitalist Spirituality* (London: Repeater, 2019); Linda Heuman, "Don't Believe the Hype," *Tricycle*, October 1, 2014; Nicholas T. Van Dam et al., "Mind the Hype: A Critical Evaluation and Prescriptive Agenda for Research on Mindfulness and Meditation," *Perspectives on Psychological Science* 13, no. 1 (2018): 36–61.
14. David L. McMahan, *The Making of Buddhist Modernism* (New York: Oxford University Press, 2008), 10–16. McMahan takes inspiration for his key themes of modernity from Charles Taylor's *Sources of the Self* (Cambridge: Cambridge University Press, 1989) and Gombrich and Obeyesekere's notion of "Buddhist Protestantism."
15. Erik Braun, *The Birth of Insight: Meditation, Modern Buddhism, and the Burmese Monk Ledi Sayadaw* (Chicago: University of Chicago Press, 2013); Daniel M. Stuart, "Insight Transformed: Coming to Terms with Mindfulness in South Asian and Global Frames," *Religions of South Asia* 11, nos. 2–3 (2017): 158–81; Daniel M. Stuart, *S. N. Goenka: Emissary of Insight* (Boulder, CO: Shambhala, 2020).

16. Ann Gleig, "From Buddhist Hippies to Buddhist Geeks: The Emergence of Buddhist Postmodernism?," *Journal of Global Buddhism* 15 (2014): 22; Ann Gleig, *American Dharma: Buddhism Beyond Modernity* (New Haven, CT: Yale University Press, 2019).
17. Jay Michaelson, *Evolving Dharma: Meditation, Buddhism, and the Next Generation of Enlightenment* (Berkeley, CA: North Atlantic, 2013).
18. Daniel Ingram, "The Dharma Overground," https://www.dharmaoverground.org/dharma-wiki/-/wiki/Main/arahat.
19. "Executive Summary," EPRC, accessed August 1, 2022, https://theeprc.org/executive-summary/.
20. Buddhist Geeks (website), https://www.buddhistgeeks.org/; Vincent Horn, Twitter, February 5, 2021, 9:31 p.m.
21. John Tresch, "Buddhify Your Android," *Tricycle*, December 5, 2015, https://tricycle.org/trikedaily/buddhify-your-android/. Brasington and his associates charge only the cost needed to run the course, depending for their livelihood on *dana* (donations from grateful students).
22. MN 139.9, cited in Brasington, *Right Concentration*, 90.
23. Ganeri's account of "attentionalism" arises in dialogue with Buddhaghosa. See Jonardon Ganeri, *Attention, Not Self* (Oxford: Oxford University Press, 2017).
24. Quotes from Quli, "Multiple Buddhist Modernisms," 232–33; Ajahn Brahm, *Mindfulness, Bliss, and Beyond: A Meditator's Handbook* (New York: Simon & Schuster, 2006); Thanissaro Bhikku, trans., *The Wings to Awakening: An Anthology from the Pali Canon* (Valley Center, CA: Metta Forest Monastery, 1996), 261; Shaila Catherine, *Focused and Fearless: A Meditator's Guide to States of Deep Joy, Calm, and Clarity* (New York: Simon & Schuster, 2008).

III

ATTENTION, TECHNOLOGY, AND CULTURE

9

WEARABLE ATTENTION

Course Corrections for Wandering Minds

NATASHA DOW SCHÜLL

A SCENE OF ATTENTION: THE VIGILANT PROSTHESIS

"It is increasingly hard to be attentive given the many distractions surrounding us." These words appear over a still image of a large lecture hall in which a class is taking place. Rows of college students face their laptop screens, their professor on the stage beyond backed by her projected slide deck of equations.[1] "We have developed a unique solution: a device that a person can put on in the moments when they want to be attentive."

The promotional video (figure 9.1) takes us through a series of these moments: learning, driving, writing. In the first scenario, a young woman with a fashionable haircut wearing red lipstick and a moss-colored cardigan struggles to listen and take notes as her mathematics professor delivers a lecture at a whiteboard. She attempts to shake herself into alertness without success. "Jane is trying to stay focused on the lecture, but she feels more and more disengaged," we are told. "Thus, she decides to put on her AttentivU glasses." Jane reaches for a pair of stylish black spectacles and puts them on. Soon, we hear a chirping noise and watch as she instantly straightens her back, blinks open her

FIGURE 9.1 Stills from a promotional video for AttentivU. Nataliya Kosmyna et al., "AttentivU: A Biofeedback System for Real-Time Monitoring and Improvement of Engagement," in *CHI EA '19: Extended Abstracts of the 2019 CHI Conference on Human Factors in Computing Systems* (New York: Association for Computing Machinery, 2019), 1–2, http://doi.org/10.1145/3290607.3311768.

eyes, and picks up the pace of her note-taking; she is now alert, confident, and able to follow along with the lecture. The glasses, we infer, do not correct wearers' vision but rather course correct their attention.

TECHNOLOGIES OF ATTENTION

Over the past decade, scholars and journalists have criticized the ways that mobile telephony, email and texting, smartphone apps, social media, streaming entertainment, video games, and other gamified internet interactions have sought to monetize our attention in the so-called click economy by attracting it, hooking it, and holding it.[2] This ongoing critique has forged a strong association between digital technology and toxic psychological effects, with an emphasis on themes of distraction, addiction, and exhaustion. A social consensus has formed: although digital technologies might extend the range of human communication, expand experiential horizons, and afford new modes of self-expression, their intensified temporalities and relentless demands for sensory and cognitive engagement pull us into coercive loops of escape and self-forgetting and exhaust our capacity to resist.

The dominant response to this predicament takes the form of expert advice on how to avoid "attentional serfdom"[3] by better policing the distracting aspects of technology: dimming screens, disabling or filtering notifications, installing ad blockers, instituting no-phone rules at home, using smart timers and automatic cutoffs,[4] and enrolling in a range of "detox" programs.[5] Relatively little notice has been given to the new crop of digital technologies that have sprung up in the wake of their more toxic kin, with promises to help beleaguered humans not by blocking, weakening, or dampening the stimuli that distract them but by supporting, strengthening, and cultivating their attentional capacities. This genre of technology runs the gamut from computer software designed to track, plan, and guide how time is spent to wearable course-correcting devices such as the AttentivU glasses described in the opening scene. Forming a distinct segment of the so-called digital health industry,[6] they present digital alternatives to the mix of drugs, tools, and techniques humans have long used to modulate their attention, including coffee, pills, sensory stimulation (e.g., light, noise, and cold), and practices such as meditation, chanting, and patterned bodily movement. In today's hyper-capitalist societies in which attention is deemed to be in crisis, most of these methods fall within the range of admissible interventions one might experiment with in the classroom, the office, the trading floor, the factory, or the gym—settings where human performance depends on the capacity for sustained focus on the part of individual agents. As technological fixes to an alleged technological problem, digital devices that sense and stimulate brain activity might seem an especially dubious class of remedies—yet, as we will see, they are presented to consumers as healthier and more effective than attention-management mainstays like caffeine and popular prescription drugs.

My aim in this chapter is not to argue that wearable attention technology is a poison or a cure, or that it is more one than the other, or even that, in pharmakon-like fashion, it functions as both at once.[7] Setting aside questions of how we might esteem or condemn this technology, I seek clues to its possibilities and limits by examining three devices designed to digitally mediate our attention: the Muse headband, the AttentivU glasses, and the FeelZing patch. As a cultural anthropologist, I am less concerned with the accuracy of the product designers' understandings of human attention (i.e., whether these understandings accord with the neural reality of attention) or the efficacy of their remediations (i.e., whether they "work") than with three lines of inquiry: (1) According to what conception of attention does each device operate? (2) How do its designers configure the attentional interventions it makes? (3) What kind of attentional subjects do its features and functions address, enable, and perhaps bring into being?

As will become clear, much is common across the trio of devices under discussion. Not only do they stand together in opposition to digital distraction tech, but they also stand apart from the wide range of consumer products devised to protect against distracting environmental signals by diminishing their salience (e.g., ad blockers, earplugs, noise-canceling headphones, and blinder-type headgear for open-office settings). Taking a different tack, they fortify what psychologists and neuroscientists characterize as "voluntary attention," in which subjects direct their attention to an object or task in accordance with their conscious intentions—as opposed to "involuntary attention," in which subjects react to salient signals in their environment.[8] At the same time, and in keeping with the increasingly automated terms of life, all three treat voluntary attention and its modulation as phenomena that can be delegated to technological algorithms and functions.

Yet despite their shared commitment to enhancing attentional sovereignty[9] via technological support, each device entails distinct conceptions of attention, modes of technological intervention, and models of the subject. Their respective formulas for attentional management, considered together and in contrast, reveal tensions and fissures in the dominant attentional logics of the day, suggesting that the "attentional regime," to use Yves Citton's term, is not yet settled.[10]

MUSE: THE TRAINER

Seated on the stage of a packed convention center ballroom in Las Vegas, a petite brunette named Ariel Garten, a neuroscientist and the designer and co-creator of Muse, describes the brain-sensing headband: "It's a focused-attention training tool. It helps you better manage internal and external distractions and keep your mind from wandering. It improves your attention, allows you to work better in an open-office workspace."[11] Unlike the AttentivU glasses, the Muse is worn in *advance* of attentional challenges, as a way to prepare for moments in life when attention is needed.

Giving a digital, cyborgian twist to centuries of meditation techniques, Muse allows users to access the intimate signals of their own brains via real-time audio feedback, "making the intangible tangible," to quote the company's frequent promotional refrain. The system, a simple band of rubber that runs across the forehead and is held in place behind each ear, contains four electroencephalography (EEG) sensors[12] that work to detect brain wave activity, which it communicates to wearers as the sounds of weather. An *active*, distracted, wandering mind is heard as stormy weather; a mind in its *neutral*, resting state is represented by a modest amount of wind, indicating that "attention isn't fluctuating, but there is also no deep focus present"; the wind dies down entirely with deep focus, and when one manages to hold this focus for more than five seconds, the sound of birds chirping is heard.[13] Like the biofeedback gadgets of an earlier era, Muse brings users to an awareness of their own internal processes so that they can learn to shift and control them; Garten refers to neurofeedback as "touching your own internal state."[14]

In a recorded demonstration of Muse, Garten pulls up a graph of a recent meditation session on her smartphone, tracing the ups and downs with a finger (figure 9.2): "This is what my

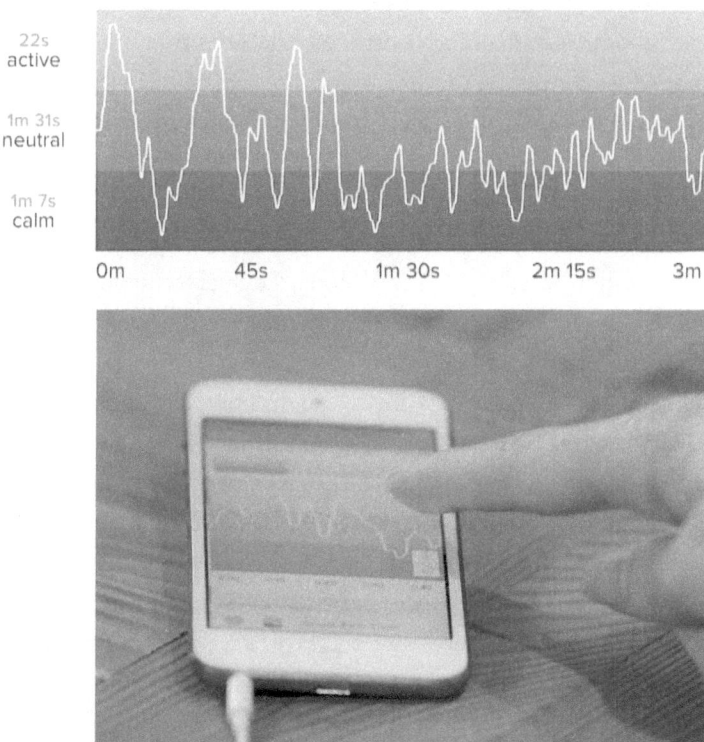

FIGURE 9.2 Screenshot from a promotional video for the Muse headband. Muse, "Muse with Co-founder Ariel Garten," YouTube video, 2:31, uploaded May 17, 2016, https://www.youtube.com/watch?v=GQlKaXguD2Y.

brain was doing during the course of the meditation," she says. "These peaks are where I was distracted, and these troughs are where I was able to bring it back down. When I began I was quite distracted, but then I was able to get in the groove. Then I had an annoying thought. I was able to bring it back down; it returned; and then I came out of it at the end."[15]

Garten characterizes the continual back-and-forth from focus to distraction that happens during meditation as an "attentional loop," noting that "each time you go through this loop, your brain is better able to recognize when you are distracted and to practice regaining control of your attention." She claims that Muse reinforces this learning process by allowing users not only to hear but also see their attentional loops so that they can "notice, more quickly and more accurately, whenever your mind is wandering."

"We constantly get distracted, and we constantly need to bring our attention back," says Chris Aimone, the chief technology officer and co-creator of Muse. He suggests that we think of attention as the "spotlight" of the mind, emphasizing our power over its movement: "We can direct this spotlight however we choose; we can control the direction of that spotlight, and we can also control the aperture . . . we can choose how we open up or close down our focus."[16] Switching metaphors, he casts Muse in the role of a personal trainer leading the user through repeated attentional loops to strengthen the "muscle for attention" in the manner of a "brain curl." "Just as exercise is necessary to train the body to fight against a sedentary lifestyle, meditation acts as a necessary tool to train the mind and become more effective in fighting off distraction," Aimone explains.[17] Meditation is "an act of will."

Yet the technology is by no means passive. If meditation is a will-based, distraction-fighting tool, then Muse is a tool *for* that tool. Rather than triggering or bestowing an attentional state, the device is designed to support our intention to direct and sustain attention—what Aimone calls "focus endurance"—by conveying our own brain signals to us in a form we can understand. Garten notes that in traditional meditation (i.e., unassisted by digital technology), minutes might pass before we notice our attention wandering and know to "bring it back"; during a session, we may experience this loop five to seven times. In contrast, during a Muse-assisted session, "you know right away, as it's happening. You know within five hundred milliseconds that you need to bring your focus back." Because Muse detects mind-wandering so quickly and allows wearers to hear it happening in real time, the device increases opportunities to practice attention control by tenfold. "Muse speeds up the attentional loop," explains Aimone. "Each time you go through this loop, you are rewiring and strengthening your brain's ability to be aware of and control its own attention."

Digitally speeding up our meditation training, the makers of Muse suggest, is necessary to cope with the sped-up pace and intensified attentional demands of life today. From the documentary-style advertisement that follows "how three real Muse users use Muse in their everyday lives" to find balance amid the pressured rhythms of work, childcare, and self-care, to the ad for "The Power of Attention," an online "master course" for creative executives that begins by acknowledging that they are "constantly being pulled in multiple directions by competing priorities" (figure 9.3), the device's promotional materials promise

FIGURE 9.3 Website ad for "The Power of Attention". InteraXon Inc., "Anchors and Attentional Loops," ChooseMuse.com, 2019, https://choosemuse.com/the-power-of-attention/.

that regular sessions with Muse will allow users to more calmly and productively direct their attention where it needs to go.

ATTENTIVU: THE SUPERVISOR

The third scenario presented in the promotional video for AttentivU that opened this chapter features the protagonist Jane—played by the lead designer of AttentivU, Nataliya Kos'myna—seated at a desk working on a laptop as her smartphone repeatedly lights up and pings her with audible notifications. Flustered, she again dons the glasses, presses a button on the frame, and returns her gaze to her laptop; her phone goes silent, and she is able to sustain her concentration. "She uses her AttentivU glasses to disable notifications while she is in a state of focus," the video explains. Later, when Jane leans back to take a break, the glasses sense from her posture and relaxed eye movement that she is now "welcome to distractions" and release the stored pings waiting on her phone, through which she happily scrolls.

In effect, the glasses take the place of attention itself, as Kos'myna defined it for an audience of Microsoft engineers in 2019, citing state-of-the-art cognitive neuroscience: "a mechanism that alleviates computational burden by prioritizing processing of that subset of [sensory] information deemed to be of the highest relevance to the organism's goals."[18] Today, overloaded as it is by "the magnitude of information surrounding us, with around-the-clock internet access, and the constant shifting between increasingly complex tasks," the mechanism of attention requires scaffolding.[19] The glasses, conceived at the MIT Media Lab in 2018, were designed to provide this scaffold.

A diagram of the glasses indicates the placement of sensors that have been calibrated to track eye movement and brain

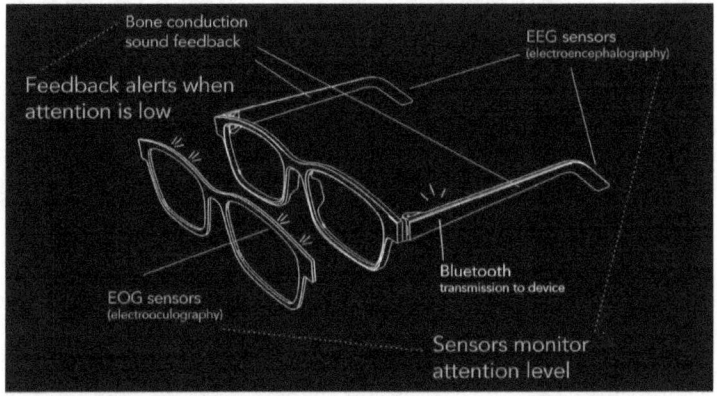

FIGURE 9.4 Diagram illustrating the features of the AttentivU glasses. Nataliya Kosmyna et al., "AttentivU: A Biofeedback System for Real-Time Monitoring and Improvement of Engagement," in *CHI EA '19: Extended Abstracts of the 2019 CHI Conference on Human Factors in Computing Systems* (New York: Association for Computing Machinery, 2019), 1–2, http://doi.org/10.1145/3290607.3311768.

activity (figure 9.4). Combined, the two data streams compute a user's cognitive load, fatigue, and focus. When the system detects that attention is wandering, it "provides real-time, subtle, haptic or audio feedback to nudge the person to become attentive again," the point being to "redirect the attention . . . to the task at hand."[20] Auditory feedback—in the form of a chirping sound that startles (rather than rewards, as in the case of Muse's chirping birds)—is delivered via a bone conduction mechanism embedded in the arms of the glasses, while an optional, unobtrusive pin clipped to clothing near the wearer's skin provides haptic feedback in the form of vibration.[21]

The glasses accompany wearers into the world (or "into the wild" as Kos'myna's lab puts it), much like glasses worn when someone wishes to see better; they are "used in the moment, in the context where sustained attention is necessary."[22] In an early experiment with a prototype, MIT students wore the glasses while attending a lecture featuring slides ranging along a continuum from boring to engaging. A control group was simply monitored while the rest of the subjects were given either targeted or random feedback in the form of auditory and haptic nudges. It was found that those provided targeted feedback when AttentivU detected their minds wandering were best able to focus on the lecture and recall its content at a later time.

AttentivU is one of several projects in the portfolio of MIT's Fluid Interfaces research group, which seeks to counter the negative effects of "today's pervasive digital devices" by developing wearable technology to enhance and supplement natural cognitive abilities—such as "motivation, attention, memory, creativity, and emotion regulation"—that are crucial to "leading a successful life."[23] In a presentation titled "From Distraction to Augmentation," the director, Pattie Maes, emphasized that her group is not interested in *imposing* the AttentivU technology on workers or students so as to surveil or discipline them but rather in *empowering* them as agents of their own behavior; the glasses are "a new type of wearable that a person could *choose* to put on when they want to be attentive and not mind-wander."[24]

Kos'myna related her own experience using the technology in her lab: when interns interrupted her focus to ask for a signature, the system picked up "a huge movement from internal to external attention" and buzzed her to get back on track. To prevent these disruptive intrusions, she now keeps a lamp on her desk that is integrated with the glasses; when she is concentrating, it glows red, a public broadcast of her brain state that warns others to keep their distance. The device runs interference for Kos'myna, insulating her fragile focus from dangerous breaks. But when she isn't attempting to stay in a focused state, she removes the glasses. "Take them off when you go to lunch," she recommends. "You don't need to be attentive during lunch. *You don't want to be attentive at all times.*"[25] Unlike the Muse designers, who understand attention as something that defines *all* human experience, Kos'myna treats it as a limited resource to be expended only in certain, prescribed situations.

Another key difference between the creators of Muse and AttentivU are the course-correcting roles they have assigned to their products. While the makers of Muse consider users to be the chief agents of attention sensing and redirecting, Kos'myna and her team delegate a greater portion of sensing and redirecting agency to the device, which stands poised to snap wandering minds back to attention. Although AttentivU wearers participate in the human–machine loop as the source of tracked data and the receiver of algorithmic nudges, their only space of action is that of a *reaction*. Recall Jane's sudden alertness in reaction to the buzzes and chirps that the glasses surreptitiously delivered during the soporific math lecture. There is no suggestion in the AttentivU marketing materials that users might employ the glasses as a training device in the mode of Muse, eventually learning to regulate their attention on their own. On the contrary, the marketing materials emphasize that "no 'special training'" is required and that users should "simply wear the device in situations where they might need it."[26] Unlike the self-actualizing subject of neurofeedback, the user of AttentivU is approached as a subject of need, reliant on the guidance of the device in situations that call for focus.

FEELZING: THE FIXER

The design logic of the third device I consider is primarily pharmacological, departing from the training logic of Muse and the prosthetic-sentinel logic of AttentivU.

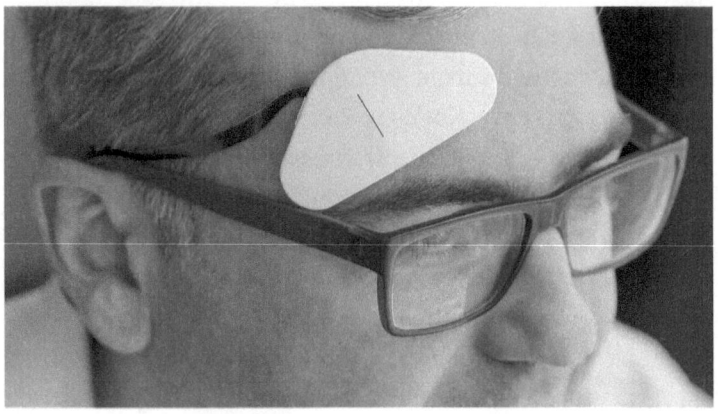

FIGURE 9.5 "I was able to use Thync instead of that second cup of coffee to boost my energy in the afternoon." Geoffrey A. Fowler, "This Gadget Gives You a Low-Voltage Pick-Me-Up," *Wall Street Journal*, July 21, 2015, https://www.wsj.com/articles/this-gadget-gives-you-a-low-voltage-pick-me-up-1437503825.

Sitting at his desk in the offices of the *Wall Street Journal* in 2015, the personal-tech columnist Geoffrey Fowler shares that he has entered an afternoon slump and is losing focus. Instead of drinking another cup of coffee, he will clear his mental fog and gain clarity by using the Thync device (figure 9.5). In his demonstration video, Fowler affixes the white triangular device to his temple, a lead running to an adhesive disc just behind his ear. Next he opens the Thync app on his phone and chooses the "energize" program, described by the company as stronger than a cup of coffee or a Red Bull energy drink.[27] He explains that for the next five minutes the device will deliver a pattern of electrical pulses to stimulate his autonomic nervous system and increase the neuron firing rate in certain areas of his brain, granting him a dose-dependent duration of attentive alertness. Fowler narrates his experience of the neurostimulation: "I'm getting a little tingle behind my ear and also a little bit of heat on my forehead." Then, "I'm going to turn it up to eighty in the interest of science, just to see what happens. Oh, wow, a lot of tingle, folks! Too high. Let's go back down to seventy-two. So now I hang out with this thing for about ten minutes. It doesn't matter if I work, watch TV, or just stare off into space." After the session, his ability to focus renewed, Fowler is ready to continue his workday.

The creators of Thync have articulated its approach to attention in direct opposition to that of Muse: "Muse makes a wearable based on EEG which can read brain waves," commented Jamie Tyler, one of Thync's founders. "It does a good job at *sensing* brain activity, but we want to *do* something to it."[28] "Most of what you see in wearables today," Sumon Pal, a neuroscientist and Thync's executive director, told a journalist, "is based on sensors, and those devices give you feedback and ask you to change yourself or change your behavior in some way to get to a desired place."[29] In contrast, "our device just *does* it. We're not sensing anything; we're *activating* the right parts of your nervous system."

Whereas Muse wearers must train their attention and AttentivU wearers must react when called to attention, Thync wearers are not "in the loop" at all as conscious subjects; there are no inputs from users to calibrate the system, no nudges to bring them back on track, no graphs to reflect upon after the fact. Certainly wearers' bodies are in a loop with the device in the sense that its patterned electrical pulses have been digitally engineered with human physiology in mind. And Thync's makers take pains to emphasize that it introduces nothing foreign to the human system but rather "actually speaks the brain's natural language": electricity. "Energize using your own nervous system," suggests company advertising. "At Thync," the website proclaims, "we focus on harnessing the electrical communication system that is already innate to the human body."[30] "It's my own body's neurons and nerves that are turning on," says a user in a promotional video. Nonetheless, as when drinking a cup of coffee, following the initial decision to do so there is little to no "will" required with respect to attention cultivation. As Fowler tells us in his video, while the device is acting on him, "It doesn't matter if I work, watch TV or just stare off into space." A journalist who tried the technology concurs: "It's like choosing a workout program, only instead of doing squats or lunges, the technology does the work for you."[31]

Because users found the initial model of the technology cumbersome on account of the required pairing with a smartphone, the application of gel strips to the neck to hold the device in place, and frequent charging, it was redesigned in 2020 in the form of an easy-to-apply patch and rebranded as "FeelZing." In a promotional spot demonstrating how easy it is to use, the actor simply pulls a tab to activate the patch and affixes it behind her ear; by the end of the thirty-second video, we can see from her alert affect that she is experiencing the effects of the FeelZing (figure 9.6). The video instructs users to remove the patch after seven minutes and expect to feel focused and energized for up to four hours, noting that they can safely reuse the patch within twenty-four hours to meet various day-to-day attentional challenges, from working on the computer to putting a child to bed.[32]

At four dollars a patch (and slightly more for the extra-strength version now also available), FeelZing's attention solution "costs less than a cup of coffee" and, unlike coffee, produces "no jitters or anxiety." "I feel like I had drunk three cups of coffee and I was more focused on my work and my thinking was more clear," reads one Amazon.com review. Nor does Thync's system carry the risk of overdose, as its effects are not cumulative, unlike those of ingested substances. It is "a healthier, long-term alternative to

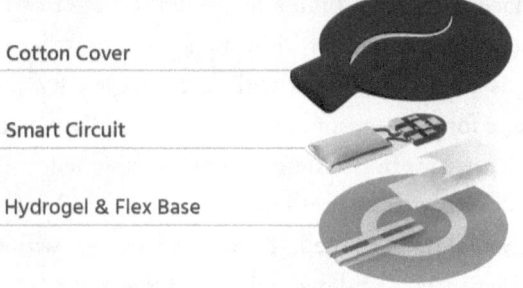

FIGURE 9.6 The FeelZing patch and its digital circuitry. "Why FeelZing Neurostimulation Is Safe" (video), FeelZing, https://feelzing.com/pages/science.

stimulants," says Dr. Jonathan Charlesworth, Thync's lead scientific adviser. Thync marketers make much of the nonchemical, "clean" nature of the patches; they promote "using electricity instead of chemicals," as one advertisement says.[33] Yet FeelZing clearly operates in a pharmaceutical fashion, delivering attention in the form of a dose of neurostimulation that lasts about four hours; it provides "a boost in energy and concentration for the moments you need: working, studying, sports, or e-sports training," as Thync's website states. Like a nicotine patch—and also like a patch in computer code—it fixes errors.

CONCLUSION: PARSING ATTENTION TECH

The three devices I have considered here sit beside one other on the shelves of Best Buy and the virtual aisles of Amazon.com, well removed from the video game systems, smartphones, and other portals of digital distraction and attention absorption on sale. They function neither to capture attention nor to insulate it from capture by blocking or dampening distractions; instead, they interface with users' brains to strengthen their capacity to pursue intended attentional aims in situations where mind-wandering is a risk. In contrast with technologies that push and pull attention for purposes of financial profit and have no concern for negative psychic effects, these devices are designed to align the behavior of users with their attentional aspirations. In this sense, all are "technologies of the self," forming an alliance with users as they struggle to attain, and maintain, a virtuous state.[34]

That the sought-after virtue is attentional sovereignty, or the ability to voluntarily direct one's focus to optimize one's life chances within contexts of productive labor—the office, the classroom, the gym, working on a laptop—should certainly give us pause. The fact that these tools of attentional well-being enable the kind of self-maximizing comportment that serves dominant economic values is, perhaps, a good reason to dismiss them alongside their more obviously toxic counterparts across the floor at Best Buy. Although the latter are designed to exploit attention and the former to cultivate it, both are invested in the same political-economic imperatives.

Yet in this chapter I have resisted collapsing them out of hand, attempting instead to discern differences in the ways that each conceptualizes attention, intervenes in it, and conceives of the attentional subject, with the hope that these differences can help us to better grasp the directions in which this technology is moving and, perhaps, what new directional possibilities might exist.

Attention: The designers of Muse conceptualize attention as an "aperture" through which we experience the world, a muscle of the mind that we can train—and, with technological guidance, train faster and better—to return from its wandering and settle where we wish. The AttentivU designers conceive of attention as a limited cognitive resource needed in select situations having to do with performance and productivity, a resource that, in today's world of intensified distraction, requires technological supervision and correction to protect it from wandering. The FeelZing designers understand attention as an internal brain state of vigilance that can, as needed, be electrically induced from the outside.

Attention intervention: The Muse headband intervenes via real-time neurofeedback that "lets you hear your own brain" as it wanders, so you can practice "bringing it back." The AttentiveU glasses also communicate audibly with wearers, but instead of transmitting the waxing and waning of focus for training purposes, they give course-correcting alerts when wandering is first detected. Thync's FeelZing patch does away with sensing altogether, delivering attention as a preformulated dose. Training, redirection, dosing. The temporalities of these interventions are likewise distinct: Muse, via short, dedicated sessions, prepares the subject to better control her attention when she is not wearing the device; AttentiveU, more prosthetic than preparatory, acts on the subject's attention only in those moments when she wears the device—which are assumed to be the only moments in which she needs to be attentive; the FeelZing patch, once affixed, sparks a state of attention in the wearer that remains in her system for a few hours, with no training or supervisory corrections involved.

The attentional subject: To be sure, all three technologies are invested—insofar as they are consumer devices operating within the same market of desires and demands—in a model of the user as one who contends daily with the appeals of digital media vying for her attention and, in the face of this onslaught, lacks the capacity to keep her mind on track in accordance with her best intentions. Following on this general model of the user as *one whose attention can be shifted* (a model in which attention-exploiting technologies are also invested), each tool takes on some portion of the attention-shifting function—but apportions and configures this function according to different understandings of the attentional subject. Muse, in coach mode, treats the wearer as an athlete-executive training for her next creative adventure; AttentivU, in overseer mode, treats her like a worker needing supervisory equipment to be kept efficiently on task; FeelZing, in pharmacological mode, treats her—whether she is an athlete, gamer, parent, or office worker—as a patient in need of medicine to sustain the sprints of attention required for successful life functioning.

The point of parsing these distinctive digital mediations has not been to identify which conception of attention is most accurate, which mode of intervention is most effective, or which model of the attentional subject is most ethical. Nor is it to deny that all of them support a neoliberal-capitalist logic for the comportment of so-called consumer sovereigns.[35] Rather, it has been to recognize, in the range of ways that individuals and their brains are scripted into loops with digital devices, that no fixed template for the technological retrofitting of attention yet exists—and also to recognize that running through these experiments in wearable attention is a strong current of ambivalence regarding attentional sovereignty itself. Indeed, each device we have examined in its own way short-circuits the very premise of attentional sovereignty by promising consumers that they can attain it by *outsourcing* it to digital functions. Whether or not human attention is in crisis, it is clear that the project of attentional sovereignty is. Perhaps the task at hand is not to recuperate but to abandon that project and the cruel optimism in which it embroils us.[36] By examining designs on attention as I have done here—alongside inquiries into the ways that individuals respond (embrace, reject, adapt, or adjust) to those designs—we might sidestep the entrenched opposition of serfdom and sovereignty and find clues to alternative forms of attending to ourselves and others.

NOTES

1. Nataliya Kosmyna et al., "AttentivU: A Biofeedback System for Real-Time Monitoring and Improvement of Engagement," in *CHI EA '19: Extended Abstracts of the 2019 CHI Conference on Human Factors in Computing Systems* (New York: Association for Computing Machinery, 2019), 1–2, http://doi.org/10.1145/3290607.3311768.
2. For a recent review of this extensive literature, see Morten Axel Pedersen, Kristoffer Albris, and Nick Seaver, "The Political Economy of Attention," *Annual Review of Anthropology* 50, no. 1 (2021): 309–25.
3. James Williams, *Stand Out of Our Light: Freedom and Resistance in the Attention Economy* (Cambridge: Cambridge University Press, 2018).
4. David M. Levy, *Mindful Tech: How to Bring Balance to Our Digital Lives* (New Haven, CT: Yale University Press, 2016); Nir Eyal and Julie Li, *Indistractable: How to Control Your Attention and Choose Your Life* (London: Bloomsbury, 2020); Wisdom 2.0, "Time Well Spent: Taking Back Our Lives and Attention—Tristan Harris, Laurie Segall," YouTube video, 22:02, uploaded March 29, 2018, https://www.youtube.com/watch?v=UJ9OqzlE _zQ; Sherry Turkle, *Alone Together: Why We Expect More from Technology and Less from Each Other* (New York: Basic Books, 2011).
5. Pepita Hesselberth, "Detox," in *Uncertain Archives: Critical Keywords for Big Data*, ed. Nanna Bonde Thylstrup et al. (Cambridge, MA: MIT Press, 2021), 141–50; Theodora Sutton, "Digital Harm and Digital

Addiction: An Anthropological View," *Anthropology Today* 36, no. 1 (2020): 17–22; Rebecca Jablonsky, "Mindbending: An Ethnography of Meditation Apps in an Age of Digital Distraction" (PhD diss., Rensselaer Polytechnic Institute, 2020).

6. Digital health technology comprises diverse devices designed to support individuals struggling to avoid what are considered the various bad habits that contribute to "lifestyle diseases" such as diabetes, heart disease, back pain, and stress. See Natasha Dow Schüll, "Data for Life: Wearable Technology and the Design of Self-Care," *BioSocieties* 11, no. 3 (2016): 317–33.

7. E.g., Bernard Stiegler, *Taking Care of Youth and the Generations*, trans. Stephen Barker (Stanford, CA: Stanford University Press, 2010); Bernard Stiegler, "Relational Ecology and the Digital Pharmakon," *Culture Machine* 13 (2012): 1–19; Natasha Dow Schüll, *Addiction by Design: Machine Gambling in Las Vegas* (Princeton, NJ: Princeton University Press, 2012).

8. Building on William James's distinction between voluntary and involuntary attention, psychologists and neuroscientists have added "covert and overt, spatial and feature, active and passive, endogenous and exogenous, top-down and bottom-up, focal and diffuse." See Carolyn Dicey Jennings, *The Attending Mind* (New York: Cambridge University Press, 2020), 23.

9. Oliver Burkeman, "Commercial Interests Exploit a Limited Resource on an Industrial Scale: Your Attention," *The Guardian*, April 1, 2015, https://www.theguardian.com/commentisfree/oliver-burkeman-column/2015/apr/01/commercial-interests-exploit-limited-resource-attention. Nick Seaver characterizes attentional sovereignty as the power "to decide what to attend to and when." Seaver, "*Homo Attentus*: Technological Backlash and the Attentional Subject," presented at the Society for Social Studies of Science Annual Meeting, New Orleans, LA, 2019.

10. Yves Citton, *The Ecology of Attention*, trans. Barnaby Norman (Cambridge: Polity, 2017).

11. Author's fieldnotes, 2015.

12. The second iteration of Muse included sensors to track heart rate, breathing, and body movements using pulse oximetry, photoplethysmography (PPG), an accelerometer, and a gyroscope.

13. Muse now offers multiple "soundscape" options to "hear" one's brain waves, including "beach," "desert," "urban," and "ambient."
14. Brenninkmeijer applies Pickering's concept of the "dance of agency" to describe the interplay of human and nonhuman actors in neurofeedback sessions, finding that the process entrains users in "a new mode of being oneself." Joanna Brenninkmeijer, *Neurotechnologies of the Self: Mind, Brain and Subjectivity* (London: Palgrave Macmillan, 2016), 145.
15. Muse, "Muse with Co-founder Ariel Garten," YouTube video, 2:31, uploaded May 17, 2016, https://www.youtube.com/watch?v=GQlKaXguD2Y.
16. InteraXon Inc., "Anchors and Attentional Loops," module of online course "The Power of Attention," accessed February 12, 2021, ChooseMuse.com, 2019, https://choosemuse.com/the-power-of-attention/.
17. InteraXon Inc., "Understanding Mental Fitness and Fatigue," module of online course "The Power of Attention," accessed February 12, 2021, ChooseMuse.com, 2019, https://choosemuse.com/the-power-of-attention/.
18. Nataliya Kosmyna, "Consumer Brain-Computer Interfaces: From Science Fiction to Reality" (presentation at the Microsoft Research Lab, Redmond, WA, March 14, 2019), https://www.media.mit.edu/events/consumer-brain-computer-interfaces-from-science-fiction-to-reality/, citing Christopher Summerfield and Tobias Egner, "Expectation (and Attention) in Visual Cognition," *Trends in Cognitive Sciences* 13, no. 9 (2009): 403–409.
19. "Project AttentivU: Overview," MIT Media Lab, https://www.media.mit.edu/projects/attentivu/overview/, accessed February 18, 2021.
20. Nataliya Kosmyna, Utkarsh Sarawgi, and Pattie Maes, "AttentivU: Evaluating the Feasibility of Biofeedback Glasses to Monitor and Improve Attention," in *Proceedings of the 2018 ACM International Joint Conference and 2018 International Symposium on Pervasive and Ubiquitous Computing and Wearable Computers* (New York: Association for Computing Machinery, 2018).
21. Elsewhere, I have examined the use of haptic technology as a method of tracking behavior that "snaps" wearers to momentary attention, or what I call "actuated attention." See Natasha D. Schüll, "HAPIfork and the Haptic Turn in Wearable Technology," in *Being Material*, ed. Marie-Pier Boucher et al. (Cambridge, MA: MIT Press, 2019), 70–75.
22. Kosmyna et al., "AttentivU."
23. "Fluid Interfaces: Overview," MIT Media Lab, https://www.media.mit.edu/groups/fluid-interfaces/overview/, accessed February 18, 2021.

24. Pattie Maes, "From Distraction to Augmentation: Technology for Optimal Performance" (presentation at the AI for a Better World: Future of Infrastructure conference, MIT Club of Northern California, San Francisco, July 14, 2020), https://www.youtube.com/watch?v=yrHTncDuQL4. The pilot version of AttentivU used an EGG headband from BrainCo's FocusEDU system. Despite Maes's insistence on the self-directed nature of the technology, the device was used in several middle school classrooms in China to track student concentration levels, which were broadcast to instructors in real time via colored lights on the forehead area of the headband: red indicated full concentration, orange was moderate concentration, and blue meant mind-wandering; see Maggie Zhang, "Self-Optimization Inside Classrooms: A Network Analysis on BrainCo Educational Neurotechnology" (BS, New York University, 2020). In this distinctly disciplinary setup, wearers were required by others to use the technology and were unaware of the signals they were transmitting. The point of using the technology was to "instill in children new codes of skilled cognitive conduct." See Ben Williamson, "Brain Data: Scanning, Scraping and Sculpting the Plastic Learning Brain Through Neurotechnology," *Postdigital Science and Education* 1 (2019): 65–86, https://doi.org/10.1007/s42438-018-0008-5.
25. Kosmyna, "Consumer Brain-Computer Interfaces."
26. Kosmyna et al., "AttentivU."
27. The original Thync device had a "calm" option said to be equivalent to a dose of oxycodone, a dose of Benadryl, or a blood alcohol content of 0.08 percent. See Geoffrey A. Fowler, "This Gadget Gives You a Low-Voltage Pick-Me-Up," *Wall Street Journal*, July 21, 2015, https://www.wsj.com/articles/this-gadget-gives-you-a-low-voltage-pick-me-up-1437503825.
28. Scott Matteson, "Tapping Into the Power of Thync," *Tech Republic* (blog), March 11, 2015, https://www.techrepublic.com/article/tapping-into-the-power-of-thync/.
29. Molly McHugh, "Heads-On with Thync, the Device That Changes Your Brain," *Daily Dot*, January 7, 2015.
30. "Electricity—The Brain and Body's Natural Language," *FeelZing Neurostimulation Blog*, https://feelzing.com/blogs/news/electricity-the-brain-and-body-s-natural-language.
31. Will Shanklin, "Hands-On: Thync Mood-Changing Wearable Is Like Doing Drugs, Without All the Bad Stuff," *New Atlas*, January 9, 2015.

32. "How to Use the Patch," FeelZing, https://feelzing.com/pages/patch, accessed January 19, 2021.
33. "FeelZing Patch—Smart Energy When You Need It," FeelZing, accessed January 19, 2021, https://feelzing.com/pages/science.
34. See Natasha Schull, "Afterword: Shifting the Terms of the Debate," *Science, Technology, & Human Values* 47, no. 2 (2021; special issue, "Shifting Attention," ed. N. Seaver, R. Jablonsky, and T. Karppi): 360–65.
35. Michel Foucault, "Technologies of the Self," in *Technologies of the Self: A Seminar with Michel Foucault*, ed. Luther H. Martin, Huck Gutman, and Patrick H. Hutton (Amherst: University of Massachusetts Press).
36. Lauren Berlant, *Cruel Optimism* (Durham, NC: Duke University Press, 2011).

10

ATTENTIONAL "OWNERSHIP"

Online Education and Self-Possession

BRIAN YUAN

A SCENE OF ATTENTION: STUDY TIME

While touring a California charter school, a journalist and the school's director peek into a ninth-grade classroom. All the students sit at laptops, absorbed in instructional videos and adaptive learning software. Although nothing seems to be happening, this is what the journalist, who is investigating the education reform movement called personalized learning, came to see. He hears clicking and the occasional murmured conversation, but the main sound in the room comes from the teacher's laptop, which is playing mellow acoustic guitar music. The school director leans over and whispers, "We put on the music because it used to get too quiet in here, and it weirded people out."[1] The object of this silent absorption is the Summit Learning Platform, an online self-paced, self-teaching system that pulls content from across the internet to replace conventional teacher-led coursework. In 2017, when this tour took place, Summit was being used at 152 schools across the United States. Since then, its footprint has more than doubled; as of fall 2021, about 400 schools teaching 80,000 students across the country are using Summit.[2]

Is the scene just described good or bad for students? Does such digital learning cultivate or impair the development of their attention? Such questions, although tempting, build on the same presuppositions as Summit Learning itself. The aim of this chapter is to analyze scenes like the one at the California charter school in terms of attentional ownership, embodiment, and differentiation. Summit Learning views students as self-owning worker-subjects who possess their own attention. Yet, in practice, the Summit Learning Platform serves as a tool of attentional differentiation. That is, Summit Learning produces technosocial environments in which students cultivate diverging modes of attentional embodiment. Analyzing this attentional embodiment in terms of harm builds on the very same underlying possessive framework by figuring attention as an unequally distributed thing or attribute.

This sort of possessive individualism has a long and sordid history in American culture, ordering subjects into hierarchies based on their ownership (or nonownership) of certain raced, gendered, and classed social properties.[3] Moreover, the possessive individual has long had attentional dimensions. Jonathan Crary argues in *Suspensions of Perception* that the modern discourse of attention emerged in the nineteenth century, pinning together the psychical substructure of the Western subject after the classical notion of the mind as impartially recording the external world had ceased being credible. In other words, the discourse of attention helped save the presuppositions of the possessive individual while offering new ways of pathologizing deficient subjects. Today, a typical example of the alignment of deficient attention with long-standing hierarchies of the human is the diagnosis of attention deficit hyperactivity disorder, which studies have shown is affected by both the race of the person being evaluated and the race of the person doing the evaluating.[4] Given this background, it is far less important to this inquiry what attention *is* than *how* attending is done. As such, this chapter is divided into two parts. The first places Summit Learning within its broader social context, then takes a close, ethnographically informed look at student accounts of their experience of the platform. Then, informed by works of phenomenological anthropology, the second part seeks to pry open the black box in which Summit's attentional differentiation happens.

ATTENDING TO SUMMIT LEARNING

Summit Learning is the work of the attention industrialist Mark Zuckerberg, whose personal foundation has donated millions of dollars and extensive engineering support to Gradient Learning, Summit's parent nonprofit. The Summit Learning Platform itself is free but comes as part of a package deal: schools get free access to the platform and specialized training, but in exchange they are expected to adopt the Summit Learning pedagogical model. This particular combination of technology and pedagogy makes Summit a leader in personalized learning. Personalized learning is a movement led by technology corporations and nonprofits with ties to Silicon Valley that centers digital technology in education reform. In a world of increasingly ubiquitous attention commodification, these alliances of education reform are a matter of some concern in their own right, but to really get a sense of the stakes of personalized learning for education and attention, it is worth quoting the definition produced by the Bill & Melinda Gates Foundation in 2014: "Personalized learning seeks to accelerate student learning by tailoring the instructional environment—what, when, how and where students learn—to address the individual needs, skills and interests of each student.

Students can take ownership of their own learning, while also developing deep, personal connections with each other, their teachers and other adults."

Here we see students being figured as possessive individuals who by rights own their education and, implicitly, their attention. Although students are sadly deprived of educational self-ownership in a conventional environment with conventional pedagogy, this definition argues, personalized learning can restore their education to them by moving teachers and other adults to the side (where they provide "personal connections") and putting the student's active, agentive fit with their "instructional environment" at center stage. The Gates definition goes on to outline a recipe for restoring educational self-ownership:

1. Competency-based progression: Each student's progress ... is continually assessed. A student advances ... as soon as he/she demonstrates mastery.
2. Flexible learning environments: Student needs drive the design of the learning environment. All operational elements ... respond and adapt to support students in achieving their goals.
3. Personal learning paths: All students are held to clear, high expectations, but each student follows a customized path that responds and adapts based on his/her individual learning progress, motivations, and goals.
4. Learner profiles: Each student has an up-to-date record of his/her individual strengths, needs, motivations, and goals.[5]

In this elaborate technosocial imaginary, motivation becomes a problem that solves itself. A student's preexisting goals, motivations, and other qualities are thoroughly and continually assessed and turned into something like a consumer profile. This profile is then turned into a customized path (presumably by being fed into a computer). As a result, students no longer need to be pushed along a curriculum. Instead, they freely and of their own accord teach themselves, valorizing their newfound educational self-proprietorship into mastery. These are attention-economy logics transplanted into teaching. Restoring students to educational self-ownership winds up meaning designing technology and pedagogy so that students' attention is no longer directed by others. Instead, they freely attend to educational content they were already motivated to attend to. Owning one's education becomes a form of owning one's attention. Teaching becomes merely a logistic problem of provisioning attentionally sovereign students with content that they *want* to attend to.

In the case of Summit, much of the heavy lifting of personalizing learning is done by the platform. Even so, the Summit pedagogical model envisions students each day alternating between self-paced self-teaching on the platform and working in small groups on platform-facilitated projects throughout the day, with self-teaching totaling up to ninety minutes a day. Aside from overseeing all this autodidacticism, teachers are to mentor each student for ten minutes a week. Naturally, theory is not a reliable guide to practice in the Summit system. Each school using Summit remains administratively independent and makes its own decisions about how to implement the system. The result tends to be that students engage in somewhat more than ninety minutes of platform time a day. For instance, a 2018 evaluation of Summit in the Indiana Area School District in Pennsylvania found that 60 percent of sixth graders reported using the platform for four or more hours a day and that about 24 percent reported using it for six hours a day—more or less the entire school day.[6] Similar experiences elsewhere are borne out by the wealth of journalistic coverage Summit has elicited in outlets large and small, along with no end of blog posts, petitions, letters to editors, comments in comment sections (variously indignant, hopeful, and mournful), and the like that together constitute an archive of on-the-ground accounts of Summit Learning.[7]

Recall our opening scene, which described a nearly silent Summit Learning classroom and what it indexes: asociality. Journalists are not alone in noting the asociality of Summit. Students and teachers both complain that the Summit model does not leave enough time for face-to-face interaction. In a characteristic comment on an article about Summit in Connecticut, a teacher tells us, "I have lost the personal connection with my students that I once built by tailoring my curriculum to their attitudes, needs, and interests. Instead, we are told to build relational capacity with students through things like shaking their hands, giving them daily affirmations, and 'launching' them each day with an inspirational quote." On the student side of the equation, we hear, "Traditional teaching is more fun and surprising. These techy programs can't really show emotion as a teacher can. Some teachers make students laugh, intentionally or unintentionally. . . . The flavor of a lesson in a classroom environment is gone, and all there is [is] a bright screen."[8]

As the transfer of knowledge is taken away from teacher–student relationships, so too is a certain desire, pleasure, and energy ("flavor"). The philosopher Bernard Stiegler argues that when "parents no longer 'know anything' and are no longer responsible for anything," a child's capacity to achieve primary identification with them is short-circuited; therefore, so too is the formation of desire.[9] Since teachers function as proxy parents, these accounts suggest that the argument holds: in an asocial teaching environment, desire leaks away.

The UI-level structure of the Summit Learning Platform does not help matters. The platform is web based, accessed through an internet browser, and students' self-directed learning units are called "focus areas." A focus area consists of objectives with associated resources, which are generally links out to the internet at large. The default Summit curriculum is available for examination online, letting us get a close look at its anatomy. Consider as symptomatic the eleventh grade "AP U.S. History Power Focus Area on Slavery in North America," which consists of twenty-four online resources:

- Eight links to the website for the 1998 PBS series *Africans in America*
- Four links to the website for the 2004 PBS series *Slavery and the Making of America*
- Two links to the website of Maryland Public Television's "Pathways to Freedom: Maryland and the Underground Railroad" educational program
- Four YouTube videos
- A link to the "Sectional Tension in the 1850s" educational unit from Khan Academy, a free online learning platform
- Four articles and a PDF produced by a smattering of other educationally oriented websites
- A broken link to the website for Colonial Williamsburg[10]

One might object to the quality of any of these resources, but that is beside the point (one might also object to the quality of material taught under more conventional conditions). Note instead that the resources are rarely stable, self-contained documents but instead point to ambiguously bounded bodies of material. In an online ecosystem designed to capture and direct attention, how is one to know what to attend *to* in a resource? Consider the relatively benign example of the *Pathways to Freedom* project. Summit links to a page titled "What Was the Underground Railroad?" Excluding a sidebar linking to other sections of the website, the page consists of three paragraphs of text, ending with the hyperlinked question, "Why was it [the Underground Railroad] called that?" The hyperlink takes one to the next page in a sequence of fifteen; the page also contains an embedded audio file called "Listen to This Passage as You Read!" The narration that follows differs from the text.[11] The student is dropped into an ambiguous landscape of educational solicitations, some of which might justifiably be read as obligate.

More troublesome examples involving clear alternative flows of material include the Khan Academy and YouTube video resources. Would it not in some sense be a desired educational outcome for a student studying slavery in America to spend hours watching whatever videos about slavery YouTube's algorithm happened to recommend? The platform that is to restore attentional ownership to students takes the work out of directing their attention by handing it over to the internet's algorithmic solicitations and persuasive design. Moreover, said solicitations and design offer the freely attending student plenty of opportunities to freely attend to extracurricular diversions. Such diversions are common in accounts of the platform. Students admit to playing online games while videos run in the background, and classroom observers note that students use YouTube to listen to music and watch videos unrelated to their coursework.[12] Recall Stiegler's liquidation of desire: the platform takes advantage of whatever *existing* desires students may have but reduces the role of teachers who might help foster *novel* desires. Understandably, students may well desire to play online games and watch amusing YouTube videos. Remote surveillance is one solution to this problem, and, despite operationalizing its shallow sort of attentional sovereignty, the Summit Learning Platform also provides tools for parents and teachers to see what students are up to. However, the aforementioned ambiguities of Summit's resources impair vision from the panopticon tower: the distinction between self-teaching and distraction is not reducible to what web page is on the screen.

Another source of attentional direction built into the Summit Learning Platform is pacing. In the style of the self-paced learning paths described in the Gates Foundation definition of *personalized learning*, classes are organized into "focus areas," "projects," "additional focus areas," and sometimes "challenge focus areas," graphically presented in the Summit Learning Platform as colored boxes in a calendrical series. Over the course of the school year, a "pacing line" moves across the series of boxes, indicating where a student should aim to be. Successfully completed areas turn green, whereas areas for which the result is only partially satisfactory turn yellow. Areas that have not been completed by the time the pacing line passes it turn red, meaning "off track."

Student reports suggest that the pacing line tends to desynchronize them from their classmates and divides them into two temporal patterns. The first is a neoliberal spin on possessive individualism, in which the offer to take ownership of one's learning and learn at one's own pace becomes a demand that expansively colonizes time and attention. A characteristic account comes from a student named Taylor who wrote to the *New York Times*: "I like how we don't have a *ton* of homework like at other schools, but I'm not sure Summit is worth it since we have more at-school work that eventually takes over our home life. It causes a lot of stress, and I often feel pressured to do extra work or have a certain focus area done. . . . Since starting Summit, I need to keep a to-do list, and I often get overly frustrated with the workload, which spills out of my 8:25 A.M.–3:15 P.M. school day" (emphasis in the original).[13] Note the terminological switch. The school no longer assigns much *home*work, but because Summit Learning is accessible at any time, from anywhere with an internet connection, at-*school* work is no longer done only at school. A good to-do list keeping, time-managing worker-student, Taylor attempts to take ownership of their learning, which leads their experiences at school and at home to converge toward a fluid, indefinite, project-based temporality.

Another student who wrote to the *New York Times* averred that "my evenings at home are spent taking notes for the endless focus areas . . . and if I don't pass the optional additional focus area, I earn a C, not [an] A. . . . Students who work faster get better grades, and therefore the students who take longer . . . get lower grades."[14] Since schools remain operationally independent, one can account for the question of grading in a variety of ways, but again we hear from an anxious student frustrated by Summit overflowing into their home life and struggling to keep up with other students going at their own (faster) speeds.

On the other side of the coin is a procrastinatory temporality in which—because a student can notionally work as slowly as they need to—the tempo of schoolwork becomes attenuated and jerks forward at year's end. For example, a student named Miette reports that "because there are no deadlines for checkpoints and such on Summit, many students feel they can complete assignments whenever they feel like it, and they aren't penalized for it."[15] Note again the terminology. Is the pacing line a deadline? To the student-worker who freely valorizes their attention into mastery, pacing amounts to a rolling deadline—but for students who fail to rise to the call, there is an alternative, equally textureless temporality in which the pacing line means that students "can finish things whenever they want as opposed to needing to manage their time."[16] This is, ironically, another version of owning one's own education and attentional capacities: the student in control of their own learning can be the student who works *as little* as they want.

THE PERILS OF SELF-POSSESSION

As should be clear from this brief sketch, Summit Learning cannot be understood with a design-it-and-they-will-attend technological determinism. The personalized learning movement is on to something: the production of a given type of student depends on how the attentional affordances of the Summit Learning Platform *fit* that student. Sometimes the reasons for these fits are difficult to account for, but there are good reasons to think that Summit, like so many contemporary attentional technologies, distributes its harms unevenly along lines of race and class. However, before diving into this we must make a detour though the work of the anthropologists Natasha Dow Schüll and Robert Desjarlais, whose ethnographic work examines how life histories, traumas, mental illnesses, and desires interact with environments and technologies to give rise to ways of being in the world.

Desjarlais, after working with people with mental illness experiencing homelessness in 1990s Boston, argued that although there is a long tradition in the West of taking experience as the unmarked ground of awareness, it has a distinctive character involving a sense of reflexive interiority (one's experience is private, internal, and reflective), hermeneutical depth (one's experience is immediate, rich, and ultimately exceeds analysis), and narrative flow (one's experience coheres and builds toward something more). He compares this with his interlocutors experiencing homelessness, who said that they "struggle along" in the face of a hail of environmental disturbances, holding themselves in an anti-introspective form of awareness focused on surface sensations in which the narrative order of time is partially suspended or ignored.[17] By demonstrating that ways of being in and perceiving the world ordinarily taken as not only ubiquitous but normatively desirable arise contingently at the intersection of the environment and an individual's body with its historically informed sensitivities, Desjarlais offers footholds for thought.

Pairing Desjarlais's work with Natasha Dow Schüll's work with people experiencing addiction to video gambling machines in early 2000s Las Vegas extends the argument technologically. In this work, Schüll observed that her interlocutors entered a "zone" in which bodily experience and human relationality were attenuated and eventually suspended by single-minded absorption into machinic interaction with a video gambling machine. Schüll noted that many of her interlocutors developed an addiction because entering the zone reduced the pain of mental illness and difficult social lives or enabled them to (paradoxically) recover a sense of control after a traumatic loss. Since the zone is what casinos are really selling, video gambling machines are scrupulously designed and redesigned with constant updates and variations to ensure that their affordances fit with the continuously developing sensitivities of gamblers.[18]

Setting out these coordinates, let us reread James Williams's now-famous essay "Stand Out of Our Light" with an eye not to attention but to an account of distraction. Williams declares that distraction dims the "spotlight" of attention on immediate tasks and goals, the "starlight" of attention on the narrative of one's life (approximately Desjarlais's narrative flow), and the "daylight" of attention on one's inner states—self-reflection—and the world (approximately Desjarlais's interiority). Put narratively, distraction "is when you sit down at a computer to fulfill all the plans you've made . . . and you find yourself forty-five minutes later having read articles . . . watched auto-playing YouTube videos about dogs . . . and . . . voyeured the life achievements of some astonishing percentage of people who are willing to publicly admit that they know you."[19]

Drifting about online, one loses track of one's goals, of what one was doing, and of oneself, and time skips a beat. Eventually one comes to in a moment of awareness. Interiority, reflexivity, and embodiment are attenuated, although not necessarily sociality, and the narrative flow of life decoheres for a while. Like the zone, this shift is the product of conscious design. Like "struggling along," it exposes the contingency of taken-for-granted, valued ways of being in the world. Bringing this approach to distraction to the case of Summit Learning, it is worth considering a point made by Jonathan Crary in *Suspensions of Perception*: "Modern distraction was not a disruption of stable or 'natural' kinds of sustained, value-laden perception that had existed for centuries but was an effect . . . of . . . attempts to produce attentiveness in human subjects."[20] Likewise, in attempting to produce subjects who own their own attention and education, Summit Learning affords students opportunities to undermine the very forms of value-laden, self-motivating attending it seeks to unleash. It should come as no surprise that when offered the affordances of the Summit Learning Platform, bored and restive students distract themselves. They do so because distraction overrides narrative flow when the narrative is boring, cuts out time when the time is felt to be empty, and suspends embodiment when one wants to be elsewhere. One might think of this way of being as a bug in the possessive individual. In its interactions with Facebook, YouTube, and the other large-scale systems of the attention economy, Summit Learning bursts open distraction as a line of flight, and the self-owning, self-directing worker-subject is leaking away through it—the attention economy liquidating its premises.

Yet, other students *do* respond to the demand to become good self-discipling, time-managing worker-subjects. What, then, makes some students more receptive to distraction? One can point to any number of factors (Stiegler, for instance, would cite the role of parents), but thinking in terms of phenomenological anthropology highlights two in particular: trauma and race. Consider the following account by a group of evaluators of a Summit Learning classroom from the Providence Public School District in Rhode Island: "Off-task student behavior was the same as, or worse than, in the more traditional classrooms, with some students observably working on assignments from other classes, viewing YouTube videos (or similar), queuing songs on playlists, toggling between Summit and entertainment websites, or pausing on work screens while chatting with neighbors."[21] These evaluators also reported that in-school violence was appallingly frequent and that many students had histories of trauma. Although these fragments can offer only a glimpse into the classroom, perhaps we can read between the lines and recall that many of Natasha Dow Schüll's interlocutors were attracted to machine gambling because the zone gave them a sense of control in the face of histories of trauma and social alienation. The practical crux of educational and attentional proprietorship is likewise greater control (or at least a sense of control) over the learning process. Perhaps the Summit Learning Platform affords students with chaotic home lives or histories of trauma opportunities to temporarily attenuate or suspend introspective, narratively cumulative ways of being in the world when they would lead to pain, anxiety, or sadness.

Providence's public schools are also overwhelmingly Black and Hispanic, and scholars of the phenomenology of race have long noted the pain of being reduced to one's skin color. Frantz Fanon famously described the experience of suddenly being transformed from a subject into one's skin as "shame. Shame and self-contempt. Nausea."[22] Acknowledging this experience, André Brock, a scholar studying Black communities online, argues that the nausea of racial objectification "is significantly reduced by the affordances [of] digital spaces."[23] This is not quite the same thing as described by the saying "On the internet nobody knows you're a dog," but inasmuch as distraction involves a certain disembodiment, it follows from Brock's argument that distraction offers some relief from patterns of discomfort and shame rooted in the phenomenology of race.

These hints at the roles of race and trauma in turning Summit Learning students into "good" or "bad" self-owning worker-subjects serve as a warning to be careful in deploying critiques of the attention economy that figure attention possessively. Even well-intended humanistic discourses that present attention as an essential (perhaps even *the* essential) characteristic of full adult personhood must be careful to avoid reifying it into an enduring *thing* belonging to individuals. It is all too easy for such a reified attention to feed into well-worn classed and raced hierarchies of the human.

NOTES

1. This account is adapted from Chris Berdik, "Tipping Point: Can Summit Put Personalized Learning Over the Top?," *Hechinger Report*, January 17, 2017, https://hechingerreport.org/tipping-point-can-summit-put-personalized-learning-top/.
2. "About Us," Summit Learning, accessed September 25, 2021, https://www.summitlearning.org/about-us.
3. Cheryl Harris, "Whiteness as Property," *Harvard Law Review* 106, no. 8 (1993): 1707–91.
4. George J. DuPaul, "Adult Ratings of Child ADHD Symptoms: Importance of Race, Role, and Context," *Journal of Abnormal Child Psychology* 48, no. 5 (2020): 673–77. See also Kess L. Ballentine, "Understanding Racial Differences in Diagnosing ODD Versus ADHD Using Critical Race Theory," *Families in Society: The Journal of Contemporary Social Services* 100, no. 3 (2019): 282–92; George J. DuPaul et al., "ADHD Parent

and Teacher Symptom Ratings: Differential Item Functioning Across Gender, Age, Race, and Ethnicity," *Journal of Abnormal Child Psychology* 48, no. 5 (2020): 679–91.
5. Bill & Melinda Gates Foundation, *A Working Definition of Personalized Learning* (Verona, NJ: New Classrooms, 2018), https://www.newclassrooms.org/wp-content/uploads/2018/08/personalized-learning-working-definition-1.pdf. It is beyond the scope of this chapter to review the history of personalized learning and its predecessors, but consider Anita Say Chan, "Venture Ed: Recycling Hype, Fixing Futures, and the Temporal Order of Edtech," in *digitalSTS: A Field Guide for Science & Technology Studies*, ed. Janet Vertesi and David Ribes (Princeton, NJ: Princeton University Press, 2019), 161–77; Heather Roberts-Mahoney, Alexander J. Means, and Mark J. Garrison, "Netflixing Human Capital Development: Personalized Learning Technology and the Corporatization of K-12 Education," *Journal of Education Policy* 31, no. 4 (2016): 405–20.
6. This is aside from the hours many students reported using Summit *outside* school. John A. Andelfinger et al., *Perceptions of the Summit Learning Platform* (Indiana: Indiana University of Pennsylvania, 2018), 163–64.
7. A methodological note: Critics of Summit Learning often deploy a rhetoric of hidden interests and individualistic motivations behind institutional structures that are difficult to substantiate and often resemble conspiracy theorizing. This rhetoric makes much of the material describing Summit that is not produced by Gradient Learning and partner institutions easy to dismiss as describing "communications issues" or "edge cases." It would be unwise to accept all claims made about Summit Learning outright, but, keeping in mind that branding statements as "conspiracy theory" or "conspiratorial" often serves to box them out of legitimate public discourse, I believe that it is important to take all materials produced about Summit *seriously* as evidence of the subjective experience of its users. In doing so, one need not decide whether any particular claim made about the system is *true*, as long as it is analyzed within a robust armature of social theory and an eye to other sources of knowledge production. On the deployment of the label *conspiracy theory*, see Jack Z. Bratich, *Conspiracy Panics: Political Rationality and Popular Culture* (Albany, NY: SUNY Press, 2008).

8. la_maestra, comment on Nick Tabor, "Mark Zuckerberg Is Trying to Transform Education. This Town Fought Back," *New York Magazine*, October 11, 2018, https://nymag.com/intelligencer/2018/10/the-connecticut-resistance-to-zucks-summit-learning-program.html; Kate Chin, quoted in The Learning Network, "What Students Are Saying About: Online Learning, Family Vacations and Moving to a New Home," *New York Times*, May 2, 2019, https://www.nytimes.com/2019/05/02/learning/what-students-are-saying-about-online-learning-family-vacations-and-moving-to-a-new-home.html.
9. Bernard Stiegler, *Taking Care of Youth and the Generations*, trans. Stephen Barker (Stanford, CA: Stanford University Press, 2010), 47.
10. "Slavery in North America," Summit Learning, accessed September 18, 2021, https://www.summitlearning.org/guest/focusareas/862825?fromCourseId=163087.
11. Maryland Public Television, "What Was the Underground Railroad?," *Pathways to Freedom: Maryland & the Underground Railroad*, accessed October 1, 2021, https://pathways.thinkport.org/about/about1.cfm.
12. Samuel Kolmar, quoted in The Learning Network, "What Students Are Saying"; Johns Hopkins Institute for Education Policy, *Providence Public School District: A Review—June 2019* (Baltimore, MD: Johns Hopkins Institute for Education Policy, June 25, 2019), https://jscholarship.library.jhu.edu/bitstream/handle/1774.2/62961/ppsd-revised-final.pdf, 31–32.
13. Taylor, quoted in The Learning Network, "What Students are Saying."
14. Emily, quoted in The Learning Network, "What Students are Saying."
15. Miette J., quoted in The Learning Network, "What Students are Saying."
16. Lukas, quoted in Colleen Faile, "Remove the Summit Personalized Learning Program from the Fairview Park City School District! Bring Education Back!," iPetitions, October 15, 2018, https://www.ipetitions.com/petition/remove-the-summit-personalized-learning-program.
17. Robert Desjarlais, "Struggling Along: The Possibilities for Experience Among the Homeless Mentally Ill," *American Anthropologist* 96, no. 4 (1994): 879. Experience has a complex history in Western philosophy that cannot be reduced to Desjarlais's account of it. See Martin Jay, *Songs of Experience: Modern American and European Variations on a*

Universal Theme (Berkeley: University of California Press, 2005). Here, Desjarlais's *experience* is taken as a specialized term describing a conventionally valued way of perceiving and encountering the world distinct from accounts of experience as bedrock or pure awareness.

18. Natasha Dow Schüll, *Addiction by Design: Machine Gambling in Las Vegas* (Princeton, NJ: Princeton University Press, 2012), 132–33.
19. James Williams, *Stand Out of Our Light: Freedom and Resistance in the Attention Economy* (Cambridge: Cambridge University Press, 2018), 49–50.
20. Jonathan Crary, *Suspensions of Perception: Attention, Spectacle, and Modern Culture* (Cambridge, MA: MIT Press, 1999), 49.
21. Johns Hopkins Institute for Education Policy, *Providence Public School District*, 32.
22. U.S. Department of Education, "Providence School District, RI, ACS 2015–2019 Profile," Institute of Education Sciences, National Center for Education Statistics, accessed September 27, 2021, https://nces.ed.gov/programs/edge/TableViewer/acsProfile/2019; Frantz Fanon, *Black Skin, White Masks* (London: Pluto, 2008), 88.
23. André Brock Jr., *Distributed Blackness: African American Cybercultures* (New York: NYU Press, 2020), 36.

11

ATTENTION IS ALL YOU NEED

Humans and Computers in the
Time of Neural Networks

NICK SEAVER

A SCENE OF ATTENTION: THE MACHINES LEARN

In 2017, a team of Google researchers designed a neural network made almost entirely out of attention.

In machine learning, *attention* refers to a particular kind of computational mechanism that tells a computer where to "look" in its input data as it generates its output. The Google team had replaced many of the conventional components of a neural network with these attention mechanisms, titling the resulting paper "Attention Is All You Need." Over the following six years, their paper would be cited more than seventy thousand times, making it one of the most cited papers in the field. And the architecture they designed, which they named the "Transformer," would be used to break records in many of machine learning's benchmark tasks.[1]

In February 2019, the machine learning company OpenAI announced that it had used the Transformer architecture to create a text-generating system it called GPT-2 (the "T" stands for "Transformer"). Given some text, like a question or the beginning of an essay, GPT-2 would produce more text, like an answer or the rest of an essay. The system had been trained on a data set of eight million websites and, as a result, was "chameleon-like," OpenAI claimed, able "to generate realistic and coherent continuations" of text in nearly any style. In fact, according to its makers, GPT-2 worked so well that they refused to release it publicly, garnering dramatic headlines like "OpenAI Builds Artificial Intelligence So Powerful It Must Be Kept Locked Up for the Good of Humanity."[2]

What made GPT-2 so threatening was its ability to produce text that was practically indistinguishable from human writing—and lots of it. When given the previous sentence as an input, GPT-2 continued the text like this:

> And all the while, the writers could produce an enormous number of those words: I feel a sudden pang of pain, a vague sense of presence, an almost magical understanding of my situation. I know precisely what it is; it is what you know, and it is important. I remember it immediately, even though I know there's a thousand things the other person doesn't know. I begin to wonder whether they are all the same, or only some, and whether or not they are all wrong. When I feel a certain level of certainty, I'm certain that there is something wrong.

GPT-2 generated uncanny text that was not always easy to interpret but that avoided many of the telltale grammatical lapses of earlier systems; it produced novel sentences that could not be traced to any particular source. Distinguishing such text from the associative style of vernacular online writing required careful attention—or was perhaps impossible.

In the wrong hands, OpenAI worried, GPT-2's plausibly human output could be used to flood social media platforms with machine-made messages, spreading misinformation, overwhelming moderators with abusive content, or swaying the apparent balance of public opinion. To use the sociologist Zeynep Tufekci's term, such a system could facilitate "denial-of-attention attacks," in which the overproduction of some speech effectively censors other speech by crowding it out. Ironically, it seemed that this assemblage of machinic attention might come to prey on the attention of humans. As one academic researcher put it, "Making all this intelligence available for commercial use allows elite political actors to control the most valuable resource possible in a democracy: our attention."[3]

Nevertheless, nine months later, OpenAI released its model, and coders promptly developed websites and software packages that allowed people with no expertise to use GPT-2 themselves—to generate passages like the excerpt given earlier, to power social media bots, or to do whatever else they might want to do with a machine constructed largely out of computational attention.

The rise of machine learning as a matter of public concern has illuminated an interesting cultural fact: attention seems to be everywhere. We find it operationalized inside computers as a calculative technique; we find it mobilized by critics of those computers as a vulnerable human faculty. Attention is a feature of individual minds, but it is also a collective capacity for deliberation. It is psychological, political, technical, and, of course, economic.

The idea that we now live in an "attention economy," in which companies are valued less for their products than for their ability to attract attention, has become commonplace. While critics worry about the creeping economization of human life, attention has come to seem naturally economic—it is limited, hence valuable, and thus a target for accumulation. Some of the most lucrative and well-known applications of machine learning are designed for just this purpose: recommender systems aim to keep users on websites, while programmatic advertisements entice users to click away from them.[4] Many of the anxieties and aspirations of contemporary life find expression in attentional terms, and machine learning has been caught up in this as well.

Attention shows up in many places, meaning many things, and it is an object of frequent debate and concern. All these qualities—the ubiquity, the polysemy, the contestation, and the explanatory capacity—are typical of the objects that the anthropologist Sherry Ortner, writing in 1973, called "key symbols." Ortner noted that many cultural orders seemed to be built around potent and central symbols: the Tibetan wheel, the Christian cross, and Nuer cattle, for instance, provided people with symbols to represent society as a whole, while also providing a common resource for analogy and explanation. Anthropologists have often (perhaps too eagerly) sought out these symbols as the "keys" to understanding cultural orders.[5]

In this chapter, I pursue the possibility that "attention," at least within the discursive world I've sketched so far, might fruitfully be understood as a key symbol, rather than simply an objective mental process or capacity. This framing un-asks some questions—Is "attention" in neural networks really attention?—and foregrounds others—How do people use the idea of attention to interpret the world and make claims about what is important to them? Thinking about attention in this way requires us to step back from the immediately accessible feelings we may have about the state of our own attention and its causes; instead, I hope to open up a space for thinking about how attention works as a catchall concept for making sense of value, selection, and human agency itself.

KNOWING WHERE TO LOOK

How did attention get into neural networks in the first place? According to most accounts in the field, it began with neural machine translation—or what is sometimes called "sequence-to-sequence learning." These machine translators are typically composed of two separate networks: an encoder and a decoder. The encoder takes the sentence to be translated and transforms it into a numerical representation, or "vector." The decoder transforms that vector into a sentence in the target language. These pairs of networks are "trained" (that is, they develop a model for turning words into numbers and back again) using large data sets of translated text.

What initially set neural machine translators apart from their predecessors was the fact that they required no intentional modeling of grammatical features or sentence structure: a whole sentence goes in, and a whole sentence comes out. For humanists, perhaps the most surprising thing about this approach is that it ever works at all. But where it fails most often, on its own terms, is in translating longer sentences. Many techniques in machine translation have this problem, tending to "forget" the beginning of a sentence by the time they reach the end or losing track of features like tense, gender, or plurality. A popular joke in the field goes like this: "What do we want?" "Natural language processing!" "When do we want it?" "Wait, what do we want again?"

Attention mechanisms were introduced as a possible solution to this problem in a 2015 conference paper titled "Neural Machine Translation by Jointly Learning to Align and Translate." The basic idea seemed simple: instead of encoding the source sentence into a single vector, the encoder would generate a separate vector for every word. These "context vectors" could indicate which other words were most relevant to a given word's meaning. As the decoder stepped through that series of vectors, translating each word in turn, it would "look" to other words as it went along: verbs could be linked to subjects, pronouns to antecedents, and so on. "Intuitively," the authors wrote, "this implements a mechanism of attention in the decoder."[6]

The term *attention* appears only briefly in the 2015 paper, in this appeal to intuition, and the paper does not formally refer to its technique as an "attention mechanism." But as these mechanisms spread, the intuitive link to attention would become a signature feature of how computer scientists explained them in papers, blog posts, and presentations. A pair of Google researchers, writing in the wake of "Attention Is All You Need," described the intuition like this: "When I'm translating a sentence, I pay special attention to the word I'm presently translating. When I'm transcribing an audio recording, I listen carefully to the segment I'm actively writing down. And if you ask me to describe the room I'm sitting in, I'll glance around at the objects I'm describing as I do so."[7]

Like human translators, machine translators would now be able to "glance around" in their source material. And this idea was not limited to translating between languages; translation had become just one instance of a more general operation that included activities like transcribing recordings and describing images. In the machinery of machine learning, these various media are treated as though they are simply different languages for representing the same underlying ideas. The numerical representations that lie between encoders and decoders are sometimes fancifully called "thought vectors," as though they represent meaning in purely ideational form, free from the specificities of any given language or medium. As one computer scientist explained to me, the goal of this work is to "decouple the meaning of a piece of data from the way it was expressed."

This broad understanding of translation is at work in an influential paper on attention mechanisms from 2016, "Show, Attend, and Tell," which presents a system for translating images into textual descriptions. "One of the most curious facets of the human visual system is the presence of attention," the authors write. "Rather than compress an entire image into a static representation, attention allows for salient features to dynamically come to the forefront as needed." Compressing an entire image into a static representation is analogous to rendering whole sentences as single vectors; attention makes description possible by allowing the visual scene to be carved into salient pieces.[8]

These explanations make plain what seems intuitively attention-like about these mechanisms, for computer scientists: they perform a kind of motivated selection, and they are contrasted with techniques that do not discriminate within their inputs. A neural machine translator that processes an entire sentence at once is not said to be paying attention, nor is a computer vision algorithm that carves its input into a uniform grid to be processed piecemeal. "Attention" appears once we have a mechanism that assesses saliency and adjusts accordingly.

Most explanations of computational attention, like the one quoted earlier, make favorable comparisons to human attention, suggesting that this similarity is what makes attention mechanisms so effective: these systems work to the extent that they emulate human mental processes. But when I interviewed the 2015 paper's first author, Dzmitry Bahdanau, in 2021, he disagreed: "If you expected the answer, 'It's because humans also pay attention to everything,' I'm not going to give you that answer," he told me. In contrast to human attention, there is no sense that computational attention is intrinsically limited. The significant constraint for a neural network is how many calculations must be done sequentially, waiting for the output of previous steps; attention mechanisms, by breaking up the input into many separate vectors, make it easier to process inputs in parallel.

What actually made computational attention so effective, as Bahdanau argued, was that it made very efficient use of parallel computing architectures. Ironically, then, attention mechanisms made it possible for a computer to pay attention to every word in a sentence at once, in a strangely dense form of attention. While individual nodes in the network may constrain their attention, they do so as part of a large attentive array, which, in principle, is unbounded. The result is a kind of attentional involution: a network full of motivated selection that can attend to the entirety of its input at the same time.

Like many concepts in artificial intelligence, then, attention has a loose connection with the human mental process that it is named for. None of the researchers I spoke with claimed that their work had any particular relevance to cognitive or neural processes, suggesting instead that "attention" was a convenient and intuitive label for a computational structure. And while some researchers pursue potential links between human and machine attention, the flexibility of this connection is a discursive resource: it can be taken more or less seriously at will.[9]

When I spoke with Kyunghyun Cho, the second author of the 2015 paper, he recounted how the term *attention* had made it into the paper in the first place: "You know, how we approached this idea had absolutely nothing to do with the concept of attention." What had happened, Cho explained, was that after the paper was initially drafted, the final author (Yoshua Bengio, a leading figure in the field of deep learning) had gone through the draft and added analogies to attention throughout. "And then I looked at it, and I was like, 'Well, I don't know what attention is, and I don't want to actually define it,'" Cho continued, "so I went through the draft once more and removed every single occurrence of 'attention,' except for a couple places where the analogy actually made sense."

In her typology of key symbols, Ortner identifies one main type as "elaborating symbols": these are "essentially analytic" in practice, used to express "relationships—parallels, isomorphisms, complementarities, and so forth—between a wide range of diverse cultural elements."[10] When computer scientists use "attention" to describe what their programs do (in the nonmathematical style they call "intuitive"), they are using it in this elaborating mode. Attention is used to draw brains and computers closer together but also to forge links between different kinds of computational processing: translation, transcription, and description all are bound more tightly together by being figured as attentional activities.

Despite the fact that attention seems to have arrived in neural networks almost by accident (at least in this paper), it has taken on a key role in how neural networks are understood and explained. The question of whether this technique is legitimately a form of attention is not significant to most of the people who work on it, and, when pressed, they do not insist on any necessary connection to human attention. Like other analogies, this one is selective, elaborating some aspects of the source symbol while neglecting others.

ATTENTION MUST BE DEFENDED

While the rise of attention in machine learning has happened largely out of public view, attention has come to occupy a central position in a growing popular critique of the technology sector.

One of the highest-profile figures in this "techlash" is a former Google employee named Tristan Harris. Harris first rose to prominence in 2013, when he still worked at Google, after an internal slide deck he made began to circulate widely among tech workers in Silicon Valley. Titled "A Call to Minimize Distraction and Respect Users' Attention," the deck argued that companies like Google had to become careful stewards of their users' minds, over which they now had unprecedented influence.

Harris took his new notoriety and founded an organization called Time Well Spent with a fellow Google employee, James Williams. They advocated for new metrics of software success, suggesting that tech companies' relentless pursuit of "engagement" (commonly measured in units of time) was producing harmful side effects, like compulsive use of social media among young people and political polarization. Williams would eventually publish a book-length manifesto, titled *Stand Out of Our Light: Freedom and Resistance in the Attention Economy*, which suggested that we might think of the market for attention as "a type of human trafficking"; by distracting us from our higher goals, attention-grabbing technology "literally dehumanizes" us.[11]

This focus on the human is typical among figures like Harris and Williams—a set of regretful former insiders whom the technology critics Ben Tarnoff and Moira Weigel have called the "new tech humanists." When Harris cofounded a successor organization to Time Well Spent, he named it the Center for Humane Technology. At a launch event in 2019, he put the central idea of new tech humanism in aphoristic form: "While we've been upgrading technology, we've been downgrading humanity." While these born-again critics have abandoned the stifling utopianism of Silicon Valley, they retain its core belief in the tremendous power of software. The problem, as they see it, is that technological advances like machine learning have been put in the service of narrow-minded goals, at humanity's expense.[12]

As with any humanism, it is instructive to look more closely at how "the human" is defined. On the Center for Humane Technology's website, we find a striking illustration (figure 11.1): the familiar depiction of the "march of progress" of human evolution, in which *Homo sapiens* stands up and walks away from its simian origins. But the image has been revised to make the human figure turn back, casting its gaze on its evolutionary heritage. Here, the human is set apart not only by stature and hair distribution but by the capacity to control its attention.

This evolutionary framing is ubiquitous among the new tech humanists, who embrace evolutionary psychology to define the human as an essentially ancient kind of animal living in a futuristic world that has escaped its control. Harris is fond of citing the sociobiologist E. O. Wilson: "We have Paleolithic emotions, medieval institutions, and godlike technology." In his book, Williams writes that software is now designed to appeal "to the

FIGURE 11.1 The march of attentional progress

lowest parts of us, to the lesser selves that our higher natures perennially struggle to overcome."[13]

Only when we are in command of our attention are we fully human, the new tech humanism suggests; when our attention has been captured by forces beyond our control, we have been dehumanized. This is a defense of what we might call "attentional sovereignty." If humans are defined by our capacity to orient our gaze, this higher capacity is always at risk, threatened by interpellation as reactive animal minds. The Center for Humane Technology warns about companies' "race to the bottom of the brain stem," an assault on their users' attentional capacities that results in "the overpowering of human nature."

New tech humanists aim to protect the proactive ability to focus that psychologists call endogenous attention. But across the many fields that have considered attention, we usually find that proactive form paired with a receptive capacity—the ability to be alerted by the outside world, or what psychologists would call exogenous attention. The new tech humanists figure this receptive capacity as a weakness, an evolutionary survival that may have been adaptive in the context of the Paleolithic savanna beloved by evolutionary psychology but which is a vulnerability in the present.

Longtime critics of the software industry have argued against these "prodigal tech bros" who, they note, still embrace the ideology that got us to this point: computer programming remains the privileged way to effect social change, and the problem with technological designs is how they are optimized, not their power itself. By directing attention toward "attention," Maya Indira Ganesh has argued, organizations like the Center for Humane Technology distract from other pressing issues like the global extraction and commodification of personal data. If the narrow worldview of Silicon Valley elites got us into this mess, then former tech workers may be the worst, not the best, guides for how to get us out.[14]

It is worth engaging with the arguments of the new tech humanists, then, not because they are particularly correct or well informed but because they have become powerful and popular. Harris has testified in front of the United States Congress, Facebook has committed to prioritizing "time well spent," and the new tech humanist critique has begun to make inroads in an industry that is notoriously unwelcoming to criticism. As large software companies begin to care more about popular backlash, they seem to be developing an understanding of their impact in terms of the new tech humanism.

Their notion of a "digital attention crisis" gathers together a host of social ills under one sign: "shortening of attention spans, polarization, outrage-ification of culture, mass narcissism, election engineering, [and] addiction to technology," as Harris has enumerated them. These problems, crossing many domains and scales of human action, are framed as the consequence of one thing: the degradation of human attention by machines. This crisis is, Harris claims, "the cultural equivalent of climate change."

Following the line of argument set out by critics like Ganesh, we might understand this foregrounding of attention as a way for objectively powerful figures to articulate a contrived marginality: the central problem of the software industry is not the automation of inequality or the extraction of resources from disempowered communities but rather a human universal that afflicts the people at the center as much as (or perhaps even more than) those without such access to power. The author and audience of Harris's first big slide deck—the mostly young, mostly white, mostly male tech workers of Silicon Valley—remain privileged agents, not only because they caused this problem, but because they experience it, too.

THE CALCULATING ENEMY

In this chapter, we have looked at two visions of attention. First, a computational attention that is selective and dense—a source of power and pattern-finding for artificial neural networks. Second, a human attention, the limits of which map out "the human" itself—a capacity to orient our gaze that some take to be our "highest self," paired with a vulnerability to distraction figured as our essential, potentially fatal, weakness. For the machine learners, attention is a formal technique, the value of which derives not from its scarcity but its shape. Attention mechanisms reorient the machinic sensorium, filling it with a new form of selection. For the new tech humanists, attention is precious, scarce, and vulnerable—the fragile seat of human agency.

These two uses of attention seem to be quite distinct. If attention in machine learning works like an elaborating symbol—a tool for analogical thinking across domains—then for the new tech humanists, attention seems to function more like Ortner's other category of keyness: as a "summarizing symbol." A summarizing symbol, as Ortner describes it, "'stands for' the system as whole," representing for the people who use it the key to their entire social order in a condensed form; thus, we might find someone claiming that the American flag, for instance, stands for some set of values they regard as distinctively American (freedom, industry, etc.), however arbitrary the link between symbol and referent may be.[15] The new tech humanist tendency to use "attention" interchangeably with concepts like "civilization," "agency," and "freedom" suggests that it functions for them as a kind of summarizing symbol, gathering together a host of social concerns under one sign.

The conflict between these two attentional domains—machine learning and human social order—is dramatized in a documentary produced by the Center for Humane Technology and released, ironically, by the video streaming platform Netflix. Amid interviews with an array of tech humanists, *The Social Dilemma* presents the story of a teenage boy developing an unhealthy relationship with his smartphone. He finds his attention pulled away from the people around him—his family, friends, and a potential love interest—by unceasing notifications and social media feeds. These temptations have been precisely engineered by a trio of anthropomorphized AI agents, who surveil the teen and scheme to capture his attention when it drifts away from his phone. The boy ends up radicalized by a fictional group called the "Extreme Center," and he is arrested at a rally, this scene being intercut with footage of actual political demonstrations.

As critics of the new tech humanism might have anticipated, the documentary does not dig into any of the structural factors that could lead to such events, instead focusing narrowly on social media as a threat to human judgment. *The Social Dilemma* is effectively antipolitical, blaming new technology for the growth of social movements, rather than the conditions those movements critique, and bemoaning "polarization" as evidence of a kind of irrationality on all sides that can be explained only through malign psychological interference. Attention operates symbolically here, as an object that might explain any disruptive social phenomenon.

While the documentary makes an argument about the central importance of human attention in contemporary, technologically mediated life, it also presents something more: the idea that being attended to by computers is dangerous. Throughout *The Social Dilemma*, the teenager is the object of unrelenting attention from the AI agents. The three figures watch a close-up video of his face; they view readouts of all his computing activity; they manipulate a holographic image of him, which hangs limp in front of an array of monitors. In the soundtrack, Nina Simone sings "I put a spell on you / because you're mine." Harris appears on-screen describing machine learning systems as containing "voodoo dolls" of their millions of users, producing uncanny effects in their targets. The racist undertones of arguments about "human downgrading" become more explicit here, as classic tropes about Black Caribbean threats to white American agency emerge to make a point about the twin dangers of being attended to and losing one's capacity to attend.[16]

So, while the new tech humanists focus explicitly on human attention, their appeal depends on a less explicit argument about computational attention. The increasing ability of computers to attend to people—to recognize their tendencies, to find relevant others to which they might be compared—is at the root of the dynamics they decry. The vision of computational attention laid out in the polemics of the new tech humanists is not precisely the same kind of attention pursued by deep learning researchers; it encompasses most uses of personal data to anticipate and shape behavior. But the two are isomorphic with each other: they are forms of motivated selection, a capacity to identify salient elements in a stream of information by referring to other parts of the stream.

Ortner suggests that her two main types of key symbol afford different kinds of analytical questions. Summarizing symbols invite us to examine "the cultural conversion of complex ideas into relatively undifferentiated commitment." We can see such a conversion at work in the discourse of the new tech humanism, whereby concerns about politics, morality, and psychological well-being are condensed into the figure of attention, understood in terms set by evolutionary psychology. Elaborating symbols, by contrast, direct us toward "questions of how thought proceeds and organizes itself through analogies, models, images, and so forth"—broadly speaking, the analysis of metaphor. Neural networks provide a clear contemporary example of the power of metaphor to interpret and shape technical activity.[17]

Summarizing and elaborating are not really two distinct types of symbol but rather kinds of symbolic function, Ortner realizes at the end of her essay. A symbol that serves as an analogical resource in one domain may be an object of reverence in another, and these uses may bleed into each other over time as people keep returning to their key symbols to make sense of the world. "Everyone knows what attention is," William James famously wrote in the early days of modern psychology, and that idea—that the significance of attention is immediately apparent and widely recognized—has allowed attention to flourish as a key symbol. Given this stature, it should perhaps come as no surprise to find attention everywhere, meaning many things to many people: a precious resource, an obvious analogy, the key to social cohesion itself.

NOTES

1. Ashish Vaswani et al., "Attention Is All You Need," in *Advances in Neural Information Processing Systems* 30, ed. Ulrike von Luxburg et al. (Red Hook, NY: Curran Associates, 2017), 5999–6009.
2. Alec Radford et al., "Better Language Models and Their Implications," OpenAI, February 14, 2019, https://openai.com/blog/better-language-models/.
3. Philip N. Howard, *Lie Machines: How to Save Democracy from Troll Armies, Deceitful Robots, Junk News Operations, and Political Operatives* (New Haven, CT: Yale University Press, 2020), 18. On "denial-of-attention attacks," see Zeynep Tufekci, *Twitter and Tear Gas: The Power and Fragility of Networked Protest* (New Haven, CT: Yale University Press, 2017).

4. The classic statement about the economic nature of attention can be found in Herbert Simon, "Designing Organizations for an Information-Rich World," in *Computers, Communications, and the Public Interest*, ed. Martin Greenberger (Baltimore, MD: Johns Hopkins University Press, 1971), 37. On recommender systems as enticements, see Nick Seaver, "Captivating Algorithms: Recommender Systems as Traps," *Journal of Material Culture* 24, no. 4 (2019): 421–36. On the economy of programmatic advertising, see Tim Hwang, *Subprime Attention Crisis: Advertising and the Time Bomb at the Heart of the Internet* (New York: FSG Originals, 2020).
5. Sherry B. Ortner, "On Key Symbols," *American Anthropologist* 75, no. 5 (1973): 1338–46.
6. Dzmitry Bahdanau, Kyunghyun Cho, and Yoshua Bengio, "Neural Machine Translation by Jointly Learning to Align and Translate," ArXiv:1409.0473 [Cs, Stat], May 19, 2016: 4, https://doi.org/10.48550/arXiv.1409.0473.
7. Chris Olah and Shan Carter, "Attention and Augmented Recurrent Neural Networks," *Distill*, September 8, 2016, https://distill.pub/2016/augmented-rnns/.
8. Kelvin Xu et al., "Show, Attend and Tell: Neural Image Caption Generation with Visual Attention," ArXiv:1502.03044 [Cs], April 19, 2016, https://doi.org/10.48550/arXiv.1502.03044.
9. For a review of efforts to link human and machine attention more explicitly, see Grace W. Lindsay, "Attention in Psychology, Neuroscience, and Machine Learning," *Frontiers in Computational Neuroscience* 14 (2020): 21.
10. Ortner, "On Key Symbols," 1343.
11. James Williams, *Stand Out of Our Light: Freedom and Resistance in the Attention Economy* (Cambridge: Cambridge University Press, 2018), 89, 80.
12. Ben Tarnoff and Moira Weigel, "Why Silicon Valley Can't Fix Itself," *The Guardian*, May 3, 2018, https://www.theguardian.com/news/2018/may/03/why-silicon-valley-cant-fix-itself-tech-humanism. See the website of the Center for Humane Technology for a recording of the 2019 launch event: https://www.humanetech.com/news/newagenda.
13. Williams, *Stand Out of Our Light*, xi.

14. On "prodigal tech bros," see Maria Farrell, "The Prodigal Techbro," *The Conversationalist* (blog), March 5, 2020, https://conversationalist.org/2020/03/05/the-prodigal-techbro/. On the inadequacy of these repentant insiders, see Lilly Irani and Rumman Chowdhury, "To Really 'Disrupt,' Tech Needs to Listen to Actual Researchers," *Wired*, June 26, 2019, https://www.wired.com/story/tech-needs-to-listen-to-actual-researchers/. On "attention" as a distraction from pressing political concerns, see Maya Indira Ganesh, "The Center for Humane Technology Doesn't Want Your Attention," *Cyborgology* (blog), February 9, 2018, https://thesocietypages.org/cyborgology/2018/02/09/the-center-for-humane-technology-doesnt-want-your-attention/.
15. Ortner, "On Key Symbols," 1340.
16. This renders computers what the cyberneticist Norbert Wiener would call "Manichaean devils," anticipating and attending to people. See Norbert Wiener, *The Human Use of Human Beings* (London: Free Association, 1989): 34–37. For more on this "cybernetic devilry," including the notion of the "calculating enemy," see Peter Galison, "The Ontology of the Enemy: Norbert Wiener and the Cybernetic Vision," *Critical Inquiry* 21, no. 1 (1994): 228–66. On the history of White American voodoo panic, see Adam Michael McGee, "Imagined Voodoo: Terror, Sex, and Racism in American Popular Culture" (PhD diss., Harvard University, 2014).
17. Ortner, "On Key Symbols," 1344.

12

MEDIUM FOCUS

JOANNA FIDUCCIA

When the composer John Cage arrived at Robert Morris's apartment in 1961, the younger artist knew exactly what he wanted to show his visitor: a sealed wooden box, as similar as anything in his studio to Cage's anti-compositional works. The box, just shy of ten inches long on each side, concealed a cassette player that could be triggered by a wire routed through its base (figure 12.1). Cage pulled up a chair, Morris pressed play, and then sounds came forth: a door shutting, footsteps echoing smartly against a floor, the short susurrus of a tape measure. Morris let the cassette run for several minutes before turning it off and announcing to his visitor, "That's it," but Cage insisted that Morris let the tape play on. He sat without moving for several hours until the recording reached its end, and then, so the story goes, left without a word. Morris called this object *Box with the Sound of Its Own Making*, inhuming it in a series of avant-garde ambits—Marcel Duchamp's 1916 *With Hidden Noise* being the closest at hand—that made conceptual transparency into its own kind of hermetic container. Yet under the guise of generous attention, Cage had performed a strong misreading of Morris's box that turned the work's tautology on its head. Cage's insistence upon hearing the

250 ATTENTION, TECHNOLOGY, AND CULTURE

FIGURE 12.1 *Box with the Sound of Its Own Making*. Robert Morris, 1961, Seattle Art Museum, Seattle, WA, wood with internal speaker. Wooden cube: 9 ¾ × 9 ¾ × 9 ¾ in. (24.8 × 24.8 × 24.8 cm); overall 46 × 9 ¾ × 9 ¾ in. (116.8 × 24.8 × 24.8 cm).

Photographer: Elizabeth Mann. Gift of the Virginia and Bagley Wright Collection. © 2023 The Estate of Robert Morris/Artists Rights Society (ARS), New York.

full recording suggested that *Box* was not unlike his own 1952 composition *4'33"*: famously a "silent" work in which the musician plays nothing during three timed movements, such that all that the audience hears is potentially everything that occurs during those several minutes. Treating Morris's *Box* similarly,

FIGURE 12.1 (*cont*)

as a durational work determined by an empty structure, Cage implied that it shared *4'33"*'s "contents": a human waltz of indeterminacy and constraint that bids us to harken the incidental music of our lives.

Retrospectively, the power of this misreading is to make possible a far different *Box*, one less autotelic than self-narrating. Containing only the soundtrack of its own making, we might say that *Box* has not so much an interior as an interiority— for what is interiority but a story of our own construction?— albeit one that is drastically attenuated. An interval of the past spins continuously within the box. Cage's hyperbolic attention underscores the contingency and capaciousness of this interval. Moreover, it rejects an understanding of the work that subsumes all that is ambient and unintentional about the box's making into the artless presence of the box itself, namely, the reading that holds that one hears *just those sounds* that correspond to *just the basic steps* of making boxes. On the contrary, in the radiance of Cage's attention, Morris's *Box* begins to seem less self-evident as its soundtrack comes to look less redundant. Gradually, the hammer blows, and the hawing of Morris's saw, the sound of his own industry as well as the room tone that is the sound of industry lapsing, rise to the fore. Though constructed in the most straightforward manner, with none of the flourish of his box-making contemporaries like Richard Artschwager or H. C. Westermann, *Box* starts to seem more like a diligently worked thing. After all, the screws holding it together have been carefully camouflaged by hole plugs; its edges are plumb and square (figure 12.2). We can begin to see how Cage's attentiveness might shade into a recognition of the craft of woodworking, minimally plied by Morris. The artist's conceptual statement quiets behind the monologue of the box itself. Attention flows toward the box's construction, toward whatever is frank and well made, or, alternatively, whatever is seductive or secretive about it.

This "flow," or orientation of attention, is a central concept of woodworking itself. The craft requires a basic understanding of wood as directional, striated with bundles of fibers and ducts, the pattern of which composes the grain. Before beginning any project, woodworkers must first establish the orientation of the grain, which means recognizing not just whether it runs along or slantwise to the plank but whether it moves in one direction or another. They can determine this direction by working

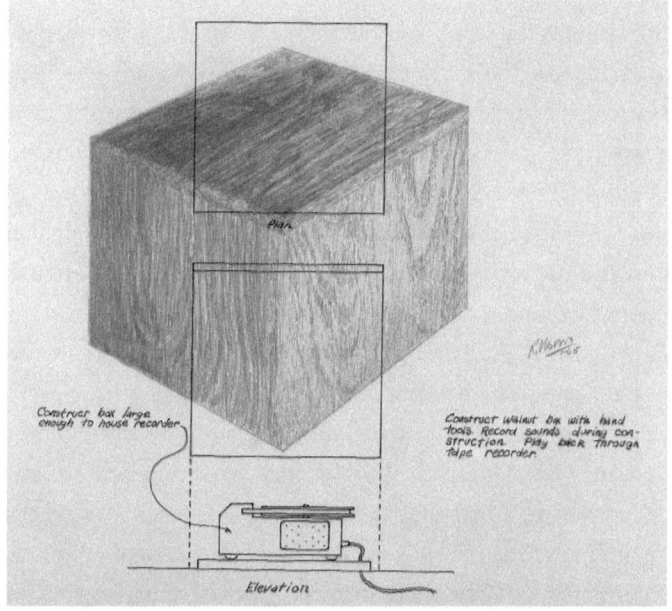

FIGURE 12.2 *Drawing for Box with the Sound of Its Own Making.* Robert Morris, 1965, Seattle Art Museum, Seattle, WA, graphite and black marker on paper and plastic. 16 ½ × 20 ½ in. Gift of the Virginia and Bagley Wright Collection.

© 2023 The Estate of Robert Morris/Artists Rights Society (ARS), New York.

the plank—uphill or downhill, in the parlance—or by observing a set of surface features. From such physiognomic scrutiny, they understand not only where a particular plank was located within the tree but how the tree itself once grew. This historically imaginative attentiveness attaches to the reified substance of time in wood as well as its extension into the future: the woodworker must *feel into* how the wood will swell, flex, and contract over time according to how it once lived as part of a tree. Moreover, orientation toward this flux in the material enforces woodworkers' own rhythms of intensity and inaction. At certain moments, they must simply wait as the wood does what it will. Attention to wood may be just this orientation to the workings of wood itself—workings that are not located exclusively in a single moment or space but that are rather ever shifting among stasis, action, and reaction, as well as present tensions and past developments.

So it may be that, at least in certain cases, material determines attention. By this, I mean not only the quality of attention but what we understand attention to be and, more importantly, how we understand the "attentive subject." This claim appears intuitive; naturally, we mobilize different senses with different intensities depending on what we're trying to observe. But it is also somewhat heretical. Philosophers who theorize attention largely seem to agree that the attentive subject is unified, even if they disagree about whether unification precedes or arises as a consequence of attention.[1] In many of these arguments, artworks hold the privileged status of objects that are both generally stable and purpose-built for attention. These assumptions are fair, if not universalizable, but when made at the exclusion of other concerns familiar to artists and art historians, they can lead to mystifying accounts of art. The role of an artist's materials is particularly consequential to any thought of attention in art history—not least because both working with and closely observing an artwork's materials frequently entail a species of attention captured by neither the hyper-focused nor the distracted subject of modernity.[2] Furthermore, this species of attention escapes the common metaphors used to describe attention: as a spotlight pointed at the world or as a filter screening its solicitations.[3] To be oriented by a material is to be, somehow, *out there* in world—not shining something onto it or selecting from it, and yet not fully absorbed in it either. We might call this state "medium focus," a term suggesting both a modality of attention and a relationship to materials.⁴ This is focus oriented by the "meeting of matter and imagination," Michael Ann Holly's definition of materiality itself.⁵ But it is also *medium*—that is, not extreme—attention, describing something that does not monopolize the mind but can be put down and taken up again. A fluctuating, discontinuous attention, it may have as much to tell us about materiality as it does about who we imagine is paying attention in the first place.

Few materials, in fact, better suggest this "meeting of matter and imagination" than wood. Take Barbara Hepworth's *Hollow Form with White* (1965): a pill-shaped monolith carved from a single piece of elm wood that has been broached by three oval apertures (figure 12.3). These recesses are painted white so that the shadows within them emphasize the double contour of each hole, echoing the annular pattern of the wood grain and its scattered burls. Elms are fast-growing trees, which made large specimens both affordable in the 1960s and challenging to carve, since their rapid expansion forces the grain to ripple out in unpredictable directions. Woodworkers call this turbulence the wood's "figure": the motif produced by the interaction of the grain with the tree's medullary rays, channels that radiate from the center of the trunk like a starburst. All wood is figured, though certain figures are more striking than others. Curly maple, for instance, has regularly alternating dark and light tones that run perpendicular to the grain, catching the light so that the surface seems to ripple like a pelt. Quilted, spalted, flamed, feathered, ray flecked, burl, and bird's eye: all terms for wood's figure, this surface apparition made from the interference of depth and plane.

The figure of Hepworth's elm is particularly disorienting, making her carved interventions more pointed. Like Romanesque groin vaults or the plaster mathematical models Hepworth

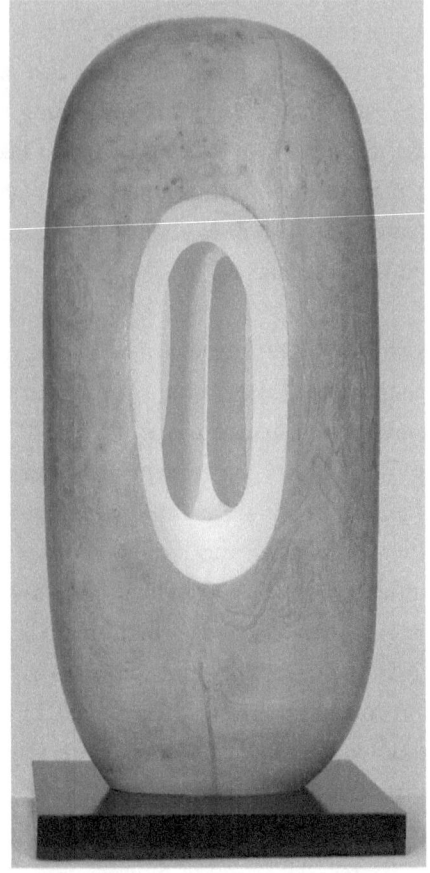

FIGURE 12.3 *Hollow Form with White.* Barbara Hepworth, 1965, Tate Britain, London, elm wood and paint. 53 × 23 × 18 ¼ in.

Photo: © Tate. © Bowness.

studied in the 1930s, her hollows are structural and orderly, in both genesis and effect the opposite of elm wood itself. Yet each of these double-contoured cells suggests a macroscopic cross-section of the elm-wood ducts, converging squarely on the arris where the other two apertures meet. This single edge appears as both a line and an advancing ridge. The inside of Hepworth's sculpture seems to project, prow-like, cutting the watery surface of the elm's figure through the very same principle that shapes it: an inner impulse in the material to drive outward.

Hollowing out *Hollow Form with White* also served a practical purpose. By excavating the center of the log, Hepworth had hoped to reduce the likelihood that it would split over time.[6] Once felled, a tree begins to undergo a process called seasoning as water trapped in its vessels evaporates. Wood that lies toward the circumference of the tree, or sapwood, has a far larger water content than the heartwood that lies closer to its center, which generally means that sapwood contracts more dramatically as it seasons. This differential contraction is the common culprit behind the checks and warping of a cut plank. Further, even seasoned timber will continue to change shape when the same cells that lost their fluids swell again in a moist environment, a characteristic called hygroscopicity. More surprisingly, seasoned timber may also warp after the woodworker has made a cut, as fibers once pulled taut by their neighbors slacken or contract as soon as they are set free, as if remembering their function when part of a living tree. In anticipation of this movement, woodworkers will cut planks larger than ultimately desired along the radial or tangential dimension, or leave joints slightly open so that the wood can expand within them. As Michael Baxandall observed, "All wood carvings remain slow-motion mobiles."[7] Hepworth's excavation, however, does not just accommodate wood's mobility and porosity; it allies those properties to a specific experience of embodiment. Like the punctured forms found across her oeuvre, the holey monolith of *Hollow Form* suggests a body that is both "less and more than a unity," as Anne Wagner writes—a body structured around a lack rather than a fantasy of inviolability.⁸ This lack, this hole, allows the wood to move, recall, and react.

It is perhaps unsurprising that wood is so often the chosen object for writers who contest the stark distinction of materials from those who work with them. The anthropologist Tim Ingold sees this separation as the legacy of an Aristotelian hylomorphic model of creation in which active minds imposed forms on passive matter. That model goes hand in glove with the elevation of technology over *technê*, the architect over the builder. Ingold proposes an alternative: the "textility of making," a responsive process of "becoming-form" that substitutes the architect's blueprint for the slicing and binding of pliant materials. He aims to evoke in his phrase less "textile" than the Sanskrit origins of *technê*: "axe," *Tasha*, and "carpenter," *taken*. In a textilic model of making, one labors foremost by following. "Practitioners," writes Ingold, "are wanderers, wayfarers, whose skill lies in their ability to *find the grain of the world's becoming* and to follow its course while bending it to their evolving purpose."⁹ Because of its flux and reactivity, wood grain stands in for all flows or forces that the skilled know to follow. But perhaps just as notable in Ingold's textilic model is the axe work, the disjunctive and labile quality of both materials and the attention that follows them. We see this disjuncture and dynamic impulse formally encoded in Hepworth's *Hollow Form*.

Some years later, Morris appeared to find himself on the same page. "A static icon-object no longer has much relevance," he announced. "The attention given to both matter and its inseparableness from the process of change is not an emphasis on the phenomenon of means. What is revealed is that art itself is an activity of change, of disorientation and shift, of violent discontinuity and mutability, of the willingness for confusion even in the service of discovering new perceptual modes."¹⁰ In other words, being oriented to process is not the same as being engrossed in materializing a form. Rather, it is an experience of disorienting "discontinuity and mutability," with the promise or payoff of "new perceptual modes." But to get there requires recognizing the confusion of forces both within and required to change a material—not a forgetting of the self but a curious involvement of it. Where, exactly, is attention given?

One resource for thinking about the location of attention is the name art history now most often associates with the topic: Michael Fried. In the late 1960s, Fried developed a theory of modernism in an essay on minimalism titled "Art and Objecthood." Although it would take full form in book-length studies of eighteenth- and nineteenth-century art over the next decades—his 1980 *Absorption and Theatricality: Painting and the Beholder in the Age of Diderot* and his 1990 *Courbet's Realism*—the theory was developed out of his rejection of Morris and other minimalists of his generation. Fried referred to "Art and Objecthood" in letters as his "sculpture-theater essay," synopsizing his central criticism of minimalism: that its objects were antithetical to art because they made recourse to a theatrical address to an individual viewer, isolating her in her own self-conscious experience as if she had encountered the work in the heightened, irreal scenario of a performance—and thus foreclosing on the community of meaning upon which art both formally and morally depends. Fried reasoned that the whole interest of minimalist objects (*interest* being these objects' shoddy placeholder for *significance*) lay in how the viewer beheld them: in endless variations of an encounter that stood in stark contrast to the absorbing instantaneity or "presentness" offered by the great modernist artworks.[11] He tracked this contrast back to an eighteenth-century tactic at the birth of modern art: the fictional negation of the beholder. When individuals in an artwork appear unaware of the viewer because they are absorbed in their own worlds, the beholder's act of viewing can remain unselfconscious. She can thus enter a "perfect trance of involvement."[12]

By contrast, "theatrical" compositions, in which figures appear posed for the benefit of a beholder, have an alienating effect. Fried takes particular umbrage at the hollowness of Morris's and Donald Judd's objects. Hollowness implies that these forms have "an inner, even secret, life," even if they appear to be simple boxes.[13] He concludes that minimalist objects are thus covertly anthropomorphic, set up in the gallery to address their audience as if they were actors on a stage, playing out a new scenario with each new visitor. Confronted with forms that seem to behold her, the beholder can no longer access the truth and conviction she felt most keenly when she had forgotten herself. Paradoxically, attending to our attention to an artwork is the surest way to break its spell.

Fried's model of attention is as much an observation about the impact of Enlightenment subjectivity on art history as an effort to contain an artwork's capacity to destabilize the ideal attentive subject. For Fried, minimalist objects involved the viewer not in a play of fixed positions (negated or absorbed, addressed or performing) but instead in "indeterminate, open-ended—and inexacting—relation."[14] Such a form of unfettered relationality replaced the communion of aesthetic meaning-making with a community founded in free-floating empathic attention.[15]

To see beyond these alternatives—free-floating or disorienting attention on one hand and fixed positions on the other—we must look further into the past, to an art historian less often invoked on the theme of attention. Between writing the two parts of his *Spätrömische Kunstindustrie*, a sweeping Hegelian enterprise that precipitated general principles of style out of the historical relations between viewers and artworks, the Austrian art historian Aloïs Riegl wrote an essay dealing more directly with the status of the beholder. His object was a fairly niche genre of painting: the Dutch group portrait, which depicts fictionalized gatherings of guilds or civic societies. In these portraits, individuals are portrayed in all their physiognomic particularity, ranged close together and yet ensconced in their own time and space, as if lodged in a dark putty. Riegl's theory was that the unity of these group portraits could not be sought on the canvas alone but instead relied on the beholder, who connects the many gazes of the depicted men like ribbons to a maypole. The Dutch group portrait, writes Riegl, "contruct[s] a bridge between figures through the representation of a selfless psychological element (attention), by means of which the individual psyches were forged together as a whole in the consciousness of the beholding subject."[16] Rather than the artwork unifying the beholder by drawing her consciousness together in the act of perception, it is the act of perception itself that unifies the painting.

Although Dutch group portraits were an outmoded genre by the time Riegl wrote his essay at the turn of the century, their capacity to address the psychology and metaphysics of attention was timely. Riegl's essay engaged with the fin-de-siècle German vogue for a Schopenhauerian "participation in the world soul" through selfless beholding, as well as the work of the psychologist Wilhelm Wundt, who sought to rescue the mind from mechanistic blank-slate models by positing attention as that which gives consciousness its distinctiveness and coherence.[17] For Riegl, however, these notions did not fully account for the significance of attention within the Dutch group portrait. For when we look at these portraits, he reasoned, we are not just the attentive beholder; we are also beheld. The group portrait suggests an attempt to reckon with the reciprocity of attention through the frontality and, indeed, theatricality of these paintings. All portraits minimally presume the existence of other individuals whom we cannot fully assimilate into our subjective experience. They therefore present test cases for the dualism of object and subject that occupied thinkers in the baroque, who coupled an interest in subjective experience with an unwillingness to sacrifice all sense of an object's independent existence. The condition Fried calls "absorption" in modern art resembles one means to circumvent that dualism.[18] This circumvention is both a classic and modern impulse, as Riegl explains: "Classical antiquity avoided [the direct address of the viewer], for it recognized only objects. Modern art can likewise dispense with it, but for the opposite reason. It recognizes only the subject, for according to its view the so-called objects are entirely reduced to perceptions of the subject."[19] The baroque, however, wished to grant the conjunction of objective and subjective phenomena for both aesthetic and ethical reasons.[20] It aimed to model a mode of mutuality that could come about only through the individualization of a psychological connection—namely, attentiveness. With too much objectivity, there could be no empathy; with too much subjectivity, there could be no collectivity.

The hero of Riegl's study is Rembrandt, whom he calls "the most distinguished practitioner of the 'painting of attention.'" Rembrandt emerges as most-Dutch-of-the-Dutch in his commitment to reconciling subjectivism with objective form, having developed a solution of incorporating Latinate pictorial strategies (in particular, the formal and narrative subordination of multiple figures to a single element or focal point) into the direct address of the beholder. Riegl contrasts the "pure, selfless attention" of the typical individual in the Dutch group portrait with the psychologically complex expressions of Rembrandt's figures, who are depicted in actual gathering spaces (a surgeon's theater, a guild's chamber) in the middle of some actual common purpose and whose nuanced gestures have "translated into physical movement . . . expressions of will and feeling." Because the attention of each individual is particularized, varying in intensity and quality, and because the faces lack the cool detachment of those in more conventional group portraits, we glimpse in these figures Rembrandt's great goal as a portraitist: "the interfusion of souls with one another and with the soul of the beholder."[21]

The acme of Rembrandt's struggle with the group portrait is *The Night Watch* (1642), in which the beholder is addressed not directly by the gaze of any figure in the painting but by the beholder's dawning awareness—as she tracks the slow stirring preparations of the troupe, such as the captain issuing an order and the lieutenant's facial expression as he hears it—that the militia is about move (figure 12.4). More to the point, that it is about to advance into the open space in front of the painting, where, inevitably, it will encounter the beholder herself. The members of the night watch don't orient their gaze toward the beholder (in fact, they don't look at her all); rather, the group collects within itself a certain (individuated) attention that will imminently flow in her direction. The problem, however, is that to arrive at this realization, the beholder must observe the painting so intensely and at such length that she is likely to dissolve the artwork into her inner experience of it (too much subjectivity!).

Rembrandt ultimately finds a resolution years later, long after the Dutch have ceased to care much for these group portraits, with *De Staalmeesters* (1662), a group portrait of the directors and chair of a textile company (figure 12.5). In the painting, the directors turn to face the beholder, each wholly particularized in body and affect, while the chair turns slightly to his right in the direction of one of his company, his hand resting palm up on the

FIGURE 12.4 *Militia Company of District II Under the Command of Captain Frans Banninck Cocq* (known as *The Night Watch*). Rembrandt van Rijn, 1642, Rijksmuseum, Amsterdam, oil on canvas. 379.5 × 435.5 cm.

FIGURE 12.5 *De Staalmeesters* (*The Sampling Officials*). Rembrandt van Rijn, 1662, Rijksmuseum, Amsterdam, oil on canvas. 191.5 × 279 cm.

table in the gesture of someone presenting a case. What sets this painting apart, Riegl argues, is the "doubling of attention": we see the directors reacting to the chair's speech, *and* we see them looking at us, as if to gauge our reaction to the chair's words. With that, we recognize that *we* are the party with whom they are negotiating. The direction of their attention is both spatial and, as in *The Night Watch*, temporal: they await our response.

Riegl cautions not to make the mistake about this direct address that Fried makes of minimalist work—that is, not to imagine that Rembrandt meant to depict the world as a momentary, subjective impression, the sense of which is wholly contained in our reaction to it. And this is where Riegl's writing on the Dutch group portrait, though concentrated on the impalpable exchange of gazes and gestures, joins his insight into Rembrandt's materials. The artist remained as committed to the reality (or objectivity) of other subjects as he did to the objectivity of external things, yet his interest in the "interfusion" of subjects and objects meant that a more conventional manner of rendering things in paintings—through crisp contours that detached them from the ground—would not suffice.[22]

Instead, Rembrandt substituted the sharp edges of haptic illusion with densely textured passages that emphasized the materiality of the paint. Rather than solicit the viewer to grasp imaginatively at the objects or faces in his paintings, his scumbled and daubed pigments intensify the surface itself. The marriage of precision and rough facture draws attention to the canvas as an interface between the expressiveness of his portraits and the substance that constitutes them.[23] It is thus not entirely surprising to find Riegl meditating on Rembrandt's tactility toward the end of an essay so focused on the gaze. Rembrandt's materials and his compositions carry out the same work: orienting the beholder to the painted image without sacrificing its separateness. Might this be another way of describing the experience of being oriented to an artwork?

The *Staalmeesters* wait upon us, introducing a peculiar temporality into the scene of attention, altogether unlike the blank eternality of contemplation. In this painting, time is something worldly and contrapuntal that catches at us as we brush against it. What is remarkable in Riegl's analysis is his conflation of this temporal unevenness with a sensual roughness, a conflation that intuits them as parallel truths about attention. Those truths seem importantly preindustrial, based in working relationships that have yet to be rigorously quantified. Following Edward Cooke, we could characterize them along the (un)measures of artisanal time, which he describes as "cumulative, partially invisible, nonlinear, and episodic."[24] Woodworking, for instance, entails intervals of not-working imposed by the material itself. So it is that the luthier trains a thin strip of wood to take on a curve by bracing it with many clamps and then waiting for it to "learn" the shape before gluing it in place. Steam bending needs "cooking time," bent lamination "setting time." Part of attentiveness to wood is an orientation to this discontinuous time, a worldly and contrapuntal responsiveness.

The notion that all attention might be discontinuous has a rich and long-running bibliography in the history of psychology.[25] Even so, it is popularly maligned as a contemporary problem, even a crisis. Attention that is not sustained, but instead bubbles up in small bursts, often gets folded into the bevy of attentional ills afflicting us today—our short attention spans, above all, which make erstwhile pleasant activities like reading a book now challenging or even painful. Such laments present discontinuous or episodic attention as a symptom of the fraying subject, as well as an inevitable consequence of the modern construction of selfhood under capitalism: once the profitable theft of attention becomes repackaged as self-expression, we begin to give it away for free, dispersing the self through the very project of social self-construction.

Craft, in turn, is mobilized as an antidote: an embodied practice that offers slow, immersive, and unbroken attention.[26] Yet this conception of craft-as-flow-state renders the craftsperson's attention more like an automatism than any form of true attentiveness.[27] What if "mastery" is not the ability to work with total immersive attention but rather the ability to navigate the relation of one's medium to time? To surf the rhythms of action and inaction demanded by the medium? Tom Martin has recently claimed that when we dwell on a craftperson's absorption in a task, or the withdrawal from consciousness of their tools or their own selves, we neglect to consider what comes to occupy their attention: the unfolding process of working itself, experienced through the relationships that arise through specific tasks.[28] The meaning of any one thing involved in this process, the craftsperson included, is bound up in its relationship to other things. In other words, the medium focus of the craftsperson, though it may at first seem to imply attention to a certain kind of thing (that is, materials), more justly describes the experience of attending to relationships.[29] This focus is characterized not by consistency and immersion but by discontinuity and change, *because no component in the relationship is expected to remain meaningfully stable*. Moreover, this focus is porous to whatever might constitute the significant relation in any one action—relations not only of tool to hand to medium but also to the craftsperson's environment, the supply chain, future users, and, in short, the "chains of obligation" that yoke work to world.[30]

These descriptions of work will likely sound familiar to anyone who has made something while having to care for something or someone else at the same time. In her 1970 essay, "A Note on the Division of Labor by Sex," Judith Brown argued that "women's work" was determined not by its suitability to certain so-called female aptitudes but by the adaptability of this work to demands already made on the historical category of women—childcare foremost among them.[31] Women's labor had to be compatible with breastfeeding and watching after young children, so it needed to be safe, repetitive, close to home, easily interrupted, and easily resumed. Spinning and weaving, for instance, met these conditions and were thus classified as women's work, a classification that produced textiles as we know them now—their techniques as well as their aesthetics and symbols. This classification also suggests the connection between these "feminine crafts" and the "textilic making" discussed through woodworking, retying the connections between the blows of the axe and the rhythmic movement of the weaving shuttle.

What is more, the discontinuous demands of care work forge subjects capable of navigating those conditions. These are subjects of attention defined neither by their autonomy nor by their capacity for immersive attention but by their obligations, by the embodied production or reproduction that makes them simultaneously "less and more than a unity." To see these subjects not as experiencing compromised attention but as exercising "medium focus" is not to say that these conditions are painless, pleasant, or even sustainable. But it is to say that these subjects' attention is nevertheless real and historically robust. Theirs is a vital model of the attentive self that has played the shadowy supporting role to what we best recognize—and today mourn—as the modern subject of attention.

NOTES

My sincere thanks to Edward Cooke for his insights into craft and woodworking and to the participants of the Princeton History of Science Workshop, in particular Hal Foster, for their generous feedback on earlier drafts of this chapter.

1. For instance, Carolyn Dicey Jennings, *The Attending Mind* (Cambridge: Cambridge University Press, 2020); Carolyn Dicey Jennings, "Attention and Perceptual Organization," *Philosophical Studies: An International Journal for Philosophy in the Analytic Tradition* 172, no. 5 (2015): 1265–78; Jesse J. Prinz, *The Conscious Brain: How Attention Engenders Experience* (Oxford: Oxford University Press, 2012); Jonardon Ganeri, *Attention, Not Self* (Oxford: Oxford University Press, 2017).
2. On the capacity of artworks to reflect critically upon these alternatives, see Jonathan Crary, *Techniques of the Observer: On Vision and Modernity in the Nineteenth Century* (Cambridge, MA: MIT Press, 1990); Jonathan Crary, *Suspensions of Perception: Attention, Spectacle, and Modern Culture* (Cambridge, MA: MIT Press, 1999); Paul North, *The Problem of Distraction* (Stanford, CA: Stanford University Press, 2011).
3. Diego Fernandez-Duque and Mark L. Johnson, "Attention Metaphors: How Metaphors Guide the Cognitive Psychology of Attention," *Cognitive Science* 23, no. 1 (1999): 83–116; Diego Fernandez-Duque and Mark L. Johnson, "Cause and Effect Theories of Attention: The Role of Conceptual Metaphors," *Review of General Psychology* 6, no 2 (2002): 152–65.
4. For the sake of sustaining the double sense of *medium*, I am allowing for some slippage between "medium" and "material," the former covering a much broader category of objects (i.e., things carved out of a material like marble, wood, or ice and those forged or cast in bronze all fall under the medium of sculpture) but a narrower range of uses than those associated with any particular material.
5. Michael Ann Holly, "Notes from the Field: Materiality," *Art Bulletin* 95, no. 1 (2013): 15.
6. Even so, seven or eight rifts opened at both the top and bottom of the sculpture, which have since been patched with filler. Dicon Nance, Hepworth's assistant, suspects that these cracks may have been caused by the white coating of paint on the inside of the hollow, which kept

the fluids from evaporating evenly along the inner faces of the wood. Matthew Gale and Chris Stephens, eds., *Barbara Hepworth: Works in the Tate Gallery Collection and the Barbara Hepworth Museum, St. Ives* (London: Tate Gallery, 1999), 232.
7. Michael Baxandall, *The Limewood Sculptors of Renaissance Germany* (New Haven, CT: Yale University Press, 1980), 36.
8. Anne Middleton Wagner, *Mother Stone: The Vitality of Modern British Sculpture* (New Haven, CT: Yale University Press, 2005), 164.
9. Tim Ingold, "The Textility of Making," *Cambridge Journal of Economics* 34 (2010): 92 (my emphasis). Ingold borrows the word *following* from Gilles Deleuze and Félix Guattari ("matter-flow can only be *followed*"), who themselves use wood as an exemplary material for being itself, structurally and visibly composed of undulating, directional fibers. Gilles Deleuze and Félix Guattari, *A Thousand Plateaus: Capitalism and Schizophrenia*, trans. Brian Massumi (London: Continuum, 2004), 451. See also Tim Ingold, "Toward an Ecology of Materials," *Annual Review of Anthropology* 41 (2012): 433. On attention as an act of following, see The Friends of Attention, *Twelve Theses on Attention*, ed. D. Graham Burnett and Stevie Knaus (Princeton, NJ: Princeton University Press, 2022); also "Twelve Theses on Attention," Friends of Attention, August 2019, https://friendsofattention.net/sites/default/files/2020-05/TWELVE-THESES-ON-ATTENTION-2019.pdf.
10. One may imagine these comments as Morris's own strong misreading of *Box*. Robert Morris, "Notes on Sculpture, Part IV: Beyond Objects," *Artforum* 7, no. 8 (1969): 54.
11. Michael Fried, "Art and Objecthood (1967)," in *Art and Objecthood* (Chicago: University of Chicago Press, 1998), 166–67.
12. Michael Fried, *Absorption and Theatricality: Painting and the Beholder in the Age of Diderot* (Berkeley: University of California Press, 1980), 103.
13. Fried, "Art and Objecthood (1967)," 156.
14. Fried, "Art and Objecthood (1967)," 155.
15. See Christa Noel Robbins's essay on Fried's phobic relationship to these de-differentiated, mobile relations, which is indebted to recent studies of minimalism that track the queerness of its formal operations. Christa Noel Robbins, "The Sensibility of Michael Fried," *Criticism* 60, no. 4 (2018): 432; David J. Getsy, *Abstract Bodies: Sixties Sculpture in the Expanded Field of Gender* (New Haven, CT: Yale University Press, 2015);

Amelia Jones, "Art History/Art Criticism: Performing Meaning," in *Performing the Body/Performing the Text*, ed. Amelia Jones and Andrew Stephenson (London: Routledge, 1999), 39–55; Jennifer Doyle and David J. Getsy, "Queer Formalisms: Jennifer Doyle and David Getsy in Conversation," *Art Journal* 72, no. 4 (2013): 58–71; Gordon Hall, "Object Lessons: Thinking Gender Variance Through Minimalist Sculpture," *Art Journal* 72, no. 4 (2013): 46–57; Tom Folland, "Robert Rauschenberg's Queer Modernism: The Early Combines and Decoration," *Art Bulletin* 92, no. 4 (2010): 348–65.

16. Aloïs Riegl, "Excerpts from *The Dutch Group Portrait (1902)*," trans. Benjamin Binstock *October* 74 (1995): 11.
17. Margaret Olin, *Forms of Representation in Alois Riegl's Theory of Art* (University Park, PA: Penn State University Press, 1992), 162.
18. Margaret Iversen notes that the internal, "even hermetically sealed," disembodied quality of Fried's absorption distinguishes absorption, both structurally and psychologically, from Riegl's model of reciprocal attention. Margaret Iversen, *Alois Riegl: Art History and Theory* (Cambridge, MA: MIT Press, 1993), 132–33.
19. Riegl and Binstock, "Excerpts," 20.
20. See Olin's powerful analysis of this essay for a sustained discussion of the ethics implicit in Riegl's theory of attention. Olin, *Forms of Representation*, 155–69.
21. Riegl and Binstock, "Excerpts," 17, 8, 9.
22. Riegl and Binstock, "Excerpts," 30.
23. Nicola Suthor has argued that Rembrandt plied this "roughness," or the emphatic surface texture of his paintings, to engage viewers both self-consciously and subliminally in the process of perception, drawing their attention to the intentional or directed nature of all perception. Nicola Suthor, *Rembrandt's Roughness* (Princeton, NJ: Princeton University Press, 2018).
24. Edward S. Cooke Jr., "Artisanal Time: Cumulative, Partially Invisible, Nonlinear, and Episodic," in *Marking Time: Objects, People, and Their Lives, 1500–1800*, eds. Edward Town and Angela McShane (New Haven, CT: Yale University Press, 2020), 83.
25. To name but two coordinates: William James's definition of *genius* not as the ability to focus solely on an idea but as the capacity to bend a thought's inevitable digressions back toward a central idea, and the far

more obscure John Perham Hylan, who hypothesized that "the power to work along the line of a certain mental activity tends to be intermittent rather than continuous," flagging and renewing with tapering force, much as a muscle fatigues. William James, *The Principles of Psychology*, vol. 1 (New York: Henry Holt and Company, 1890), 423–24; John Perham Hylan, "The Fluctuation of Attention," Series of Monograph Supplements, *Psychological Review* 2, no. 2 (1898): 2.

26. See the so-called handbook of happiness: Mihaly Csikszentmihalyi, *Flow: The Psychology of Optimal Experience* (New York: Harper & Row, 1990), as well as a more recent edited collection: Brian Bruya, *Effortless Attention: A New Perspective in the Cognitive Science of Attention and Action* (Cambridge, MA: MIT Press, 2010).

27. For a recent critique of this notion, which takes to task the association of (gendered) craft with unthinking repetition and incommunicable nonknowledge, see Michelle H. Wang, "Woven Writing in Early China," *Art History* 42, no. 5 (2019): 836–61.

28. Tom Martin, "Relational Perception and 'the Feel' for Tools in the Wooden Boat Workshop," *Phenomenology & Practice* 15, no. 2 (2020): 9.

29. Not unlike another eminent discussion of "grain." See Roland Barthes, "The Grain of the Voice," in *Image, Music, Text*, trans. Stephen Heath (New York: Hill and Wang, 1977), 179–89.

30. Cooke, "Artisanal Time," 88. The emphasis on obligation and boundedness here may suggest one means to distinguish between the discontinuity under discussion and the many pauses registered in Morris's soundtrack, which are included much as Cage's misreading intimates: as intervals of contingent, "free-floating" time.

31. In this case, we can understand "women" in Brown's argument as *just that group of individuals* who are compelled to submit to those demands. Judith K. Brown, "A Note on the Division of Labor by Sex," *American Anthropologist* 27, no. 5 (1970): 1073–78. See also Elizabeth Wayland Barber, *Women's Work: The First 20,000 Years—Women, Cloth, and Society in Early Times* (New York: Norton, 1995), 29–30.

IV
ENDGAME(S)

13

ATTENTION FAST, ATTENTION SLOW

Obsession, Compulsion, and Holding Close

YAEL GELLER

"Intoxication is a number," according to Charles Baudelaire. Digital optics is indeed a rational metaphor for intoxication, statistical intoxication, that is: a blurring of perception that affects the real as much as the figurative, as though our society were sinking into the darkness of a voluntary blindness, its will to digital power finally contaminating the horizon of sight as well as knowledge.

Paul Virilio, 1994

A SCENE OF ATTENTION: ON THE TRAIN

Imagine a train with two passengers on it. One has attention deficit disorder (ADD) and is prone to distraction. The other has obsessive-compulsive disorder (OCD) and is prone to fixation. Imagine that the train is going so fast that they experience their surroundings at very high speed. They look out of the window, all forms fleeting. The traveler with OCD fixates his gaze upon a tree that he is bound to lose; the traveler with ADD enjoys the rapidly changing vista.

DROMOLOGY AND ATTENTION

That train is not merely, or simply, a means of transportation. In a much broader sense, it is the very condition for seeing. If we count not only the technology-dependent locomotion of our bodies but also the motion of the things around us, we realize that a great deal of our seeing occurs at very high speed and that a great deal of our attentiveness is conditioned and mediated by very high speed.

We have long assumed that technology changes our perceptual modalities, but most of the time we do not really give this notion our attention. It is not the train, of course, that chiefly possesses seeing in our world of technology but rather the screen—of a smartphone, a computer, a tablet, a smart watch, or a television. But first, we must hop on a train. Already there (that is, already then), the landscape appeared only to disappear. It is not coincidental, for reasons other than those imagined before, that one of the first cinematic events, and the one that is often remembered as the first in film mythology was the Lumière brothers' *The Arrival of a Train*. To our eyes, the train and the screen are essentially the same; the difference is that in the train we move fast, whereas on the screen, the images in front of us move fast. In both cases, speed mediates vision. As speed is relative, it does not matter who is doing the moving.

We are following the *dromology* of Paul Virilio, specifically his notion of the *aesthetics of disappearance*. This idea refers to the swift exposure to forms that appear only in the instant in which they light up.[1] Things around us, Virilio says, "continue to disappear in the intense illumination of projection and diffusion."[2] Speed of transmission succeeded, hence, the speed of transportation not only in kinematic standards (the ideal of rapidity); transmission also outstripped transportation in altering our ecology.

With remarkable daring, the dromologist anticipated that this dromocratic regime would eventually lead to "a final oblivion of matter and of our own presence in the world."[3] Although daunting, this prediction does not seem far-fetched. The idea that the screen will eventually make us disappear will perhaps make more sense if you recall the last time that you dozed off in front of the television. When paying attention to an object, you tie your fate to it. If this object incessantly appears to disappear, so will you. The point is not that today we pay attention the same way as before, but to less important things and for shorter periods of time than in the past. The point is that attention is disintegrating. I take Virilio seriously, which makes my task twofold: it consists in asking how "matter" and how "our own presence" will fall into oblivion.

Now that I have asked the question, I will rephrase it: What happens to the individual when speed mediates attention? (This rephrasing assumes that *attention* somehow links "matter" to "our presence," and I will assess this assumption). This chapter orbits the question using OCD and ADD as two lesser gravitational forces. They are not two answers to the question but two phenomena with which to consider it. ADD is a natural candidate because it figures prominently in many deliberations on the effects of the screen on the mind. Obsession less so, which I find odd. I argue that the two conditions are at least equally pertinent to a discussion of the crisis of attention.

PATHOLOGIES OF ATTENTION

A long time ago, a few leading psychiatrists thought that phenomenology could be used to elucidate mental pathologies. One was Eugène Minkowski, who is credited with describing schizophrenia in terms of a disturbed temporality. Minkowski was unique among these psychiatrists in that he went beyond trying to fathom the inner world of patients through an analysis of their experience of time; he also postulated that healthy psyches have a sense of rhythm with the ambient world. More specifically, he believed that this sense of rhythm is a primary condition for mental health. Tinkering with Eugen Bleuler's famous distinction between *schizoidia* (the principle of withdrawing and returning to oneself) and *syntonia* (taking part in social life), he emphasized temporality. The experiences of a healthy subject, he believed, are accompanied by a feeling of becoming, of advancing toward a future. And crucially, this becoming can be *in sync* with the ambient world. Schizoidia, in his view, is the tendency for a private rhythm; When not sufficiently balanced, it thwarts the engagement in *ambient becoming*.[4]

Unfortunately, this kind of phenomenological ideation on psychopathology resides on the margins of contemporary psychiatry.[5] Current mainstream psychiatry is too occupied with drugs to notice that the essential project of psychiatric phenomenology is far from complete. For this reason—although it is surely not the only reason—most psychiatrists would not consider OCD as having to do with attention. The underlying assumption is that it has to do with *affect*, not with cognition; indeed, before it occupied its own chapter in the DSM-5, in the third and fourth editions of the DSMs it was included with affective disorders.[6] That is not to say that OCD does not have to do with affect (of course it does), but so does ADD. Psychiatric nosology is unstable and capricious, and one should never take it too seriously. Despite what the DSM says, affect and cognition are not separable, and both are involved in OCD and ADD. The issue with anxiety is that that is exactly where they both collapse.

Therefore, without anxiety, I will try to join what is left of that old phenomenological tradition by again taking up a temporal category—speed—to moor our contemplations. Attention takes place in space and time, as everything else does, and as with everything else, it has undergone tremendous change amid our contemporary rhythm of the ambient world. The socially shared experience of time, with which we humans must synchronize our psychic rhythmicity, is after all no longer human. If synchronization with the ambient world (in the form of a shared experience of time) is indeed a prerequisite for mental health, then it becomes very difficult not to become morbid.

For Bernard Stiegler, "to *pay attention* is essentially *to wait*."[7] And while there is an association between *attention* and the French for "to wait"—*attendre*—the origins of *attend* (meaning "to pay attention") are in the Latin *attendere*, meaning "to *stretch to*."[8] It is easy to see how *waiting* can be taken to mean stretching toward the thing for which one waits, but when considering *attention*, the emphasis should be on the gesture toward, not on the passive standing still.

Putting aside the realm of the abstract—the noetic attention we pay to concepts or emotions—we begin with the immediate *being-in-the-world*. Before the screen came along (we can leave the train for now), we primarily paid attention to things we could physically attend—things that had an appearance and were at our disposal. The objects toward which our mind stretched were almost always objects we could materially verify. Although there were some exceptions (for example, people paid attention to the stars), attending the thing to which attention was given was the dominant, taken-for-granted condition for attention. This was the unconscious assumption in front of a seed, a prey animal, a machine, a waterfall, a breast. All these things had both an appearance and an immediate, actual, material reference.

In 2018, the results of research conducted in Munich were published in *Scientific Reports*.[9] In a rather simple experiment, people were asked to estimate the length of two bars, one horizontal and the other vertical. The results showed that length was more accurately estimated visually than by touch. However, the authors also found that most of their subjects were more confident in their estimates by touch. What is interesting about this finding is not the accuracy of any estimation or the illusory confidence in any sensory input. Rather, it is that it appears that people are prone to tactile "fact-checking." You ask someone to pay attention to something in front of her, and her instinct is to touch it—to attend the thing, to feel its materiality as if doing so is the only way to complete the gesture of attention. Ample evidence supports this notion. For example, babies always reach their hands toward what catches their attention. And we can easily swap ontogenesis with phylogenesis: "Hardly any faculty," wrote Darwin, "is more important for the intellectual progress of man than the power of attention. Animals clearly manifest this power, as when a cat watches by a hole and prepares *to spring on its prey*. Wild animals sometimes become so absorbed when thus engaged, that they may be *easily approached*."[10] Whether catching or getting caught, attention ends up in maximum contact.

The promise of contact is what speed—of transportation but mostly of transmission—deprives of the body and mind. The first time my daughter saw my image on her grandfather's iPhone was quite dramatic, even heartbreaking, but her frustration was merely what we repeatedly repress. Humans have always been drawn to reflective surfaces—mirrors, glass, diamonds, ice—but the unprecedented allure of the screen is second only to its deceptiveness. You turn it on, and it becomes a facade of everything. And as dazzled as you are, you eventually forget that this everything is nothing at all. This is something of the current technological Dasein.

The train was designed to bring us closer to things, but technology does more than fulfill our desire to get close. For it to bring us closer, it needs to stop. Screens and phones are everywhere, always on, thus severing the stretching of the mind from any material anchoring. Nowadays, we are continually absorbed in things we *cannot* materially access. We talk to someone on the phone, with whom we are not communicating in the flesh. We watch a program on television, but it has no actuality for us. The things we cannot access in matter—these are the things designed to draw our attention. Further, our attention is constantly being drawn to where attending is impossible not merely in space but also in time. We watch things that happened yesterday or a year ago; we constantly watch our earlier selves. The great Israeli poet Dahlia Ravikovitch wrote, "Narcissus was so much in love with himself. Only a fool doesn't understand he loved the river, too." Lucky Narcissus. If *we* drown, we will end up steeping in dry emptiness.

For Virilio, things were always getting worse. With analog technology, the material reference of the image, its origin, was not overt but at least traceable by means of analogy. In the digital age, visual information is harvested from "reality," transformed into digital information, then again into an image that quickly disappears. This process leaves no material trace, so we are left with the ghost of matter.[11] Is there a doubting Thomas inside us who must touch the crucifixion wounds?

Alex Kanevsky stunningly captures the nature of this technologically determined sensory frustration (figure 13.1). We are so often looking at what is either moving on or vanishing that our eyes become blind to blindness. The incessant, bright, and fast

FIGURE 13.1 *Dear Friend.* Alex Kanevsky, 2018, oil on linen.

oscillations of appearance and disappearance overwhelm the eye, yet the eye never gets to see it. Kanevsky makes us aware of this assault. He simultaneously manages to make present both appearance and disappearance, thus seizing the very waning of matter.

So far, I have discussed only "matter." Next, I look at the "final oblivion of our presence," which requires a second rephrasing of my question: What happens to a person whose eyes are ceaselessly occupied by empty, flickering representations with which fleshly contact is impossible? Put another way, what happens to a person whose ears so often listen to voices of absent bodies?

On a descriptive level, one can say that the sensorium is fragmented. I was in a park in Sao Paulo, Brazil, when completing the first draft of this chapter, simultaneously smelling the local flora, hearing the voice of my partner in Israel, and looking at a video of a concert in Canada (which took place fifty years ago). In *The Vision Machine*, Virilio writes about the aesthetic of disappearance, which has "arisen from the unprecedented limits imposed on subjective vision by the *instrumental splitting* of modes of perception."[12] That's our broken sensorium. When I was sitting in the park in Sao Paulo, my faculties of perception were scattered by the devices I was using. The effect of the screen with its tiny microphone, much more than the pervasive, perpetual, perceptual overdose it permits, threatens the essence of attention. On its way to final oblivion, our presence is first divided.

In Hebrew, when you ask someone to pay attention to something, you literally ask her to put her heart onto it. You do not ask for a directing of a single sense, not even for the harmonized sensorium in its entirety. You go for the heart—the synecdochic core of being. If you want someone's attention, you want her *attendance*, and you want the ideal form of attendance: the entire sensorium with all faculties of the mind *and* body. In English, we often say, "Pay attention to me," and if the demand is not met, we say something like, "Come on, *be* with me."

Sensorial unity is severely compromised, but even that is not the whole picture. With regard to the central nervous system, electroencephalography (EEG) has shown that in front of the television, the "orienting response"—the instinctive visual or auditory reaction to any sudden stimulus—happens so frequently that the alpha brain waves that characterize relaxed wakefulness are nearly eliminated. When triggered again and again, gamma waves, which characterize intense focus or concentration, are also reduced. Paradoxically, the screen makes us more alert but less focused.[13] The orienting response is also called "involuntary attention," which I find more disturbing than ironic. The Screen, you see, activates involuntary attention with such intensity that the neurological contrivance that facilitates attention leads to a loss of focus. Attention spins into absent-mindedness. "Attention span" becomes "attention spin."

RETHINKING NORMAL

In the words of the dromologist, we now live in a "location that has no location," where instead of attending an agora, we turn on "a cathode-ray screen, where the shadows and specters of a community dance amid their processes of disappearance."[14] Virilio's strokes are big and bold, yet his phenomenological critique of speed rushes toward "chronopolitics." All we do here is patiently fill in the gaps. We are now slowing down in the field of psychopathology (or perhaps "pathonormality").[15]

Adhering to the etymology of *attention*, I will rephrase my question a final time, not to answer it but to further commit myself: What happens to us when (1) attending is no longer the thrust and support of attention; (2) the faculty of attention can no longer unify our perceptual capacities; and (3) the neurological contrivance of attention is in itself turned against attention?[16]

Let's hop on that train again. Two passengers. The traveler with ADD has a racing mind—a spinning one, easy to redirect, almost impossible to arrest. Did he adjust better than his fellow? Maybe he did . . . but to what? The waning of matter? The traveler with OCD chose an object never to let go of, putting on blinders against so much of everything else. But he will succumb to anxiety. Should we take something of his tormented subversiveness? We need to get to know these two better.

In 1987, the name for attention deficit disorder was replaced with "attention deficit *hyperactivity* disorder" (ADHD), so ADD is now considered an outdated term. The reason I chose to use ADD at the beginning of this chapter is because this label refers solely to attention. Hyperactivity is also relevant, of course, as it has to do with speed. A much earlier term used to describe the phenomenon we identify today as ADHD was *hyperkinetic impulse disorder*, which is pretty much all about speed. Historically, considerations of speed (in the sense of a pathological rhythm) with regard to this phenomenon were no less consistent—albeit much less coherent—than the consideration of attention. But that's not the only point I want to make. I mention all these labels to clarify that it does not matter whether ADD, ADHD, "mental restlessness," hyperkinesis of childhood, or any other label history can offer is indeed a "mental disorder." Many psychiatrists believe that it is a *biological* or *neurological* abnormality, whereas some see it as an imaginary diagnosis, and speakers on both sides can be quite convincing. But this quarrel is irrelevant here. There is a *cultural concept*, as well as a designated phenomenon, and that, for me, is enough. Some people, mostly children, are identified as more distracted, impulsive, hyperactive, and inattentive than others, and their situation—I am purposefully avoiding the word *disorder* here—is understood to be an issue of attention. So I'm going to stick with ADD.

The disintegration of attention is so explicitly stated with regard to ADD that it almost seems to be the locus of our infinite and very anxious projections. Everything we fear is there: distraction, absent-mindedness, impulsiveness, forgetfulness, "disorderedness," lack of control, disability, loss of presence. The criteria specified in the fifth edition of the DSM even include a tendency *to lose materials*.[17]

As mentioned earlier, obsession is rarely conceptualized as a problem of attention, but I think there is much to be gained in treating it as such. In *The Principles of Psychology*, William James says of attention that "focalization and concentration of consciousness are of its essence."[18] And the *Collins English Dictionary* defines *attention* as "the concentrated direction of the mind." What mind is more "localized and concentrated" or more "directed" upon an object than one that is obsessive? Obsession is perhaps an exaggerated form of attention, but when subtracting anxiety and assuming that the object of attention is worthy of obsession, obsession can be the epitome of attention. Of course, only in theory. Abandoning everything else for one object of attention should never be a preferred disposition.

Either way, I side with the few who think that obsession is a matter of attention; moreover, given that OCD is one of the most common psychiatric diagnoses, I believe that one might gain something by thinking of it as a revolt against speed in a capitalist society. Obsession is by nature a fixation, a refusal to move fast, and in some cases a refusal to move at all—a refusal to leave the door, the car, the stove; a refusal to detach the mind from whatever it is the mind has chosen. One of the hallmarks of the symptomatology of obsession is what is called "primary slowness," which refers to the very long time it takes for a person with an obsession to perform activities like taking a shower, dressing, or getting ready to leave the house.[19]

But there's more to it than that. Obsession is not merely a resistance to speed; it is also a hunger for sheer materiality. Hoarding things is a common comorbidity of OCD. Also, the empirical literature shows that, when wanting to make sure the door is closed or the light switch is off, individuals diagnosed with OCD touch these things. They *attend* the doorknob; they don't just look at it.

Let me summarize briefly. The coordinates I'm proposing in relation to temporality and materiality are the following. In the empire of speed, under the rule of the aesthetics of disappearance, one can identify two extremes on a spectrum: ADD, characterizing people who move fast and often, are easily sidetracked, tend to lose things, and respond well to a drug we call "speed"—and OCD, characterizing people who move slowly or not at all, hoard things, and hold on to the materiality that speed threatens.

What is in the middle? If ADD and OCD are two polemic coordinates, what is normal, in between?

CONCLUSION

For centuries, the words *obsession* and *possession* went hand in hand. Historically, in Latin, *obsessio* and *possessio* both referred to the process of besieging a city. An obsessed city was a city surrounded by the enemy but whose citadel remained intact; a possessed city was a city whose walls had been breached and was conquered. Later, the military battle became a metaphor, and the terms came to be used to describe demonic assault. The city became the person; the citadel became the soul. A possessed person was one whom the devil controlled completely. He wasn't aware of the devil's hold on him; his soul, the citadel, was conquered. The obsessed person, on the other hand, was aware of being besieged by the devil, since the devil didn't have complete control over him. In psychiatric discourse, ever since Freud, people with mental illness have been labeled as either ego dystonic or ego syntonic.

Today, it seems that *obsession* and *possession* have split. *Possession* has a legal sense, *obsession* a clinical one. But has their relationship really changed?

Eric Schmidt, the CEO of Google from 2001 to 2011, said in an interview in the late 1990s with *PCWeek* magazine that the twenty-first century would become synonymous with what he called the "attention economy" and that the most dominant global corporations would be those that succeed in maximizing the attention they drew. In the reincarnation of metaphors, it is quite clear who the devil is now. The obsessed are easy to recognize as well (actually, they are usually self-diagnosed). But whose attention is possessed? Is this the new normal?

The obsessed are on to something, but they are not quite there. They traded moving forward for hesitation, spinning in the world for spinning inward. But mostly, their slowness is accompanied with that anxiety we artificially left behind. The price is high. Can we take only a part of this? If what is at stake is indeed our presence in the world, is it possible to learn something from obsession? Can we slow down without the anxiety? Can we engage in repetition and commit ourselves to repetition, not because of doubt or intruding thoughts but with delight and courage? Can we win back materiality by insisting on attendance and then let go of material objects to reap the benefits of virtuality *without* being possessed? Our goal should be to recapture lived time and space despite the technology we have created to master them. As opposed to the success of any technology, which always eventually becomes public, the success of such an effort lies in its particularity. The challenge is great, but simple logic can still guide us: if attendance is desired, aspiring nostalgically to a slow pace is not quite the answer. One needs to know when to stop.

NOTES

The epigraph is from Paul Virilio, *The Vision Machine*, trans. Julie Rose (Bloomington: Indiana University Press, 1994), 76.

1. Paul Virilio, *The Aesthetics of Disappearance*, trans. Philip Beitchman (Los Angeles: Semiotext[e], 1991).
2. Virilio, *The Aesthetics*, 65.
3. Virilio, *The Aesthetics*, 111.
4. Eugène Minkowski, *Lived Time: Phenomenological and Psychopathological Studies*, trans. Nancy Metzel (1933; repr., Evanston, IL: Northwestern University Press, 1970).

5. Such ideation occurred mostly in the first half of the twentieth century. See Allegra R. P. Fryxell, "Psychopathologies of Time: Defining Mental Illness in Early 20th-Century Psychiatry," *History of the Human Sciences* 32, no. 2 (2019): 3–31. Josef Parnas, Louis Sass, and Thomas Fuchs are three of the few who currently work on phenomenological psychopathology.
6. American Psychiatric Association, *Diagnostic and Statistical Manual of Mental Disorders*, 3rd ed. (Washington, DC: American Psychiatric Association, 1980), 225; American Psychiatric Association, *Diagnostic and Statistical Manual of Mental Disorders*, 4th ed. (Washington, DC: American Psychiatric Association, 1994), 393.
7. Virilio, *The Vision Machine*, 76; Bernard Stiegler, *Taking Care of Youth and the Generations*, trans. Stephen Barker (Stanford, CA: Stanford University Press, 2010), 96.
8. "Attend," *Merriam-Webster.com Dictionary*, https://www.merriam-webster.com/dictionary/attend.
9. Merle T. Fairhurst et al., "Confidence Is Higher in Touch Than in Vision in Cases of Perceptual Ambiguity," *Scientific Reports* 8 (2018): 15604.
10. Charles Darwin, *The Descent of Man, and Selection in Relation to Sex* (1871; repr. Princeton, NJ: Princeton University Press, 1981), 44. My emphasis.
11. "With the progressive digitalization of audiovisual, tactile, and olfactory information going hand in glove with the decline of immediate sensations, the analogue resemblance between what is close at hand and comparable would yield primacy to the numerical probability alone of things distant—of all things distant. And would in this way pollute our sensory ecology once and for all." Paul Virilio, *The Information Bomb*, trans. Chris Turner (London: Verso, 2006), 114.
12. Virilio, *The Vision Machine*, 49.
13. Michael E. Smith and Alan Gevins, "Attention and brain Activity While Watching Television: Components of Viewer Engagement," *Media Psychology* 6, no. 3 (2004): 285–305.
14. Paul Virilio, *Lost Dimension*, trans. Daniel Moshenberg (New York: Semiotext[e], 1991), 19.
15. Although Virilio often turns to psychopathology (for example, his meditation on petit mal at the beginning of *The Aesthetics of Disappearance*), his comments on psychopathology are usually utterly saccadic.

16. Michael I. Posner, *Attention in a Social World* (New York: Oxford University Press, 2012).
17. American Psychiatric Association. *Diagnostic and Statistical Manual of Mental Disorders*, 5th ed. (Washington, DC: American Psychiatric Association, 2012), 59.
18. William James, *The Principles of Psychology*, vol. 1 (New York: Henry Holt, 1918).
19. Benjamin James Sadock, Virginia Alcott Sadock, and Pedro Ruiz, *Kaplan & Sadock's Synopsis of Psychiatry: Behavioral Sciences/Clinical Psychiatry*, 11th ed. (Philadelphia: Wolters Kluwer, 2015).

14

UNITS OF INTENSIVE CARE

Poetic Attention and the Precarious Body

LUCY ALFORD

UNIVERSITY OF VIRGINIA MEDICAL CENTER, INTENSIVE CARE UNIT, 2015

Green walls, acrid smell. The rasping sound is the breathing machine that keeps my mother alive. It sounds like breathing, like a machine breathing. Beeps indicate her pulse, which continues to power her brain. Numbers on one screen indicate her vital signs. All either too low or too high. On another screen, wavy lines indicate the movement of her mind inside the coma, with furious spikes marking the seizures that take place beneath the surface. Each seizure, the doctors say, does some damage. We will not know how much until she wakes up, if she wakes up. Her body does not move. In places, her hair has been removed to make sites for EEG sensors; in places, the remaining hair is stuck together by the adhesive glue of the sensors. My mother loves hair. She would hate this if she were here.

This is a scene of attention.

I have dropped everything to be here, sitting in this room. My siblings and I take shifts or sit together in the room. We are keeping a vigil because there is nothing else to do. Wait. Wait for something to change, for something to happen. Hours turn into days, weeks into months. We wait for a long time. Sometimes I sit in the family waiting room down the hall and look at the children's books scattered around, the torn copies of *Parenting Magazine*. Sometimes I open my laptop and look at the file of my unfinished dissertation, which is on attention in poetry and to which I cannot attend. Sometimes I go down to smoke a cigarette in the alley by the hospital dumpsters. I don't want the nurses to see me smoking, even though some of them take smoke breaks, too. A cigarette takes me seven minutes, plus time on either end for elevators and hallways.

My mother's body is inert. It is somehow much smaller than it ever has been but also enormous. It fills the room, swelled by stillness and fluids. The fluids push into her veins through a tube in her arm. In her neck, an incision for the breathing tube. In her mouth, the feeding tube.

This is a scene of attention: vigilance kept alongside a body while the life of the body hangs in a balance, regulated and pumped externally. The machines register my mother's vital signs, even though the vital signs really belong to the machines themselves: to the pumping action of the breathing machine and the drip of the fluid bag. The sound of the breathing machine and the pulse beeps are at once signs of automation and of life. In the room, sitting alongside the bed or in a chair in the corner, my own vital signs seem too loud. Breath and heartbeat. The smell of my own sweat, the taste of cafeteria coffee rancid in my mouth.

To hold vigil is to hold nothing; to keep vigil, to keep nothing. Attention to a field in which something could happen, or into which something could enter, without guarantee. It is indefinite—duration without endpoint. A waiting for a change.

UNITS OF INTENSIVE CARE ❧ 293

This is a scene of attention nested in a hive of other scenes. One room on a hall of intensive care rooms, on a floor of intensive care rooms, in a hospital of many floors, wings, and annexes. Many lives in this web of rooms hang in the balance, like the life I keep watch over. Drugs are administered. Beds are changed. Nurses, cleaning staff, cafeteria staff: their lives hang in the balance, too, striated along predictable lines of race, class, and gender, moving in and out and through the building. The building is awake all night, humming and ventilating.

Night shift, day shift. The nurses wash my mother's body. They give intensive care over long hours with their hands. They cannot see into the future or know for how long they will be washing this particular body. They can only care for it while it is sleeping here, while we are waiting for a change in the vital signs. They have no way to speak to this body, any more than I have: we watch the screens. We attend to the flesh. Massage the feet, ankles, and calves to circulate the blood. Keep the skin clean and dry. Keep the tongue wet so it does not crack. Paint the toenails.

Inside the coma, my mother's mind is moving, but it cannot speak. I wonder if she can hear, if she can feel touch, if she dreams. I am not sure whether to hope she feels something or nothing. I wonder if she has thoughts, words, memories, images. Can she choose what to focus on? If there is any choice where she is, this would be it. What happens to attention when the life of the body itself is hanging here in the half-light, in this precarious net of vital signs provided by breath, pulse, fluids, and machines? What happens to keeping a vigil when the one you wait by may already be gone?

This is the second coma. The first was long but not this long. Weeks, not months. When she woke from the first one, she had lost language. Slowly words returned but disconnected from their meanings. Her eyes showed that she wanted to speak. Frustration, anger: signs of life. The words came from somewhere else, other than meaning. I still hear them all the time:

Purple tapioca.

Purple tapioca.

A pearl a purple a purple tapioca.

Since awakening from her second coma, my mother has recovered more and more language. The seizures that tore through her brain while she was unconscious did permanent damage. Sentences are still scrambled, and short- and long-term memory have been reduced to a loose lace weave. As with others who suffer from dementia or neurological damage, there are good days and bad, good hours and bad. Sometimes there's no light in the eyes, no sense in the words, just drifts of cognitive snow, roughened by dim frustration. She can no longer walk, and the falls are scary. But she has recovered more than we could have hoped. Most miraculously, despite the physical and cognitive losses, the fierce candle of her personality (that unpinnable quality of humor and glint of eye) was not snuffed. At the same time, it is a recovery that wavers, comes in and out of focus. And because she is mortal (like me, like us), it will not last forever.

We do not last forever. And there is an ocean of precarious life between full presence and full absence. The precarity of our body-houses problematizes the line between "life" and "death" insofar as our bodies are constantly regenerating and passing away. The attentional state of this vigil—being with a beloved, without words, in this space between here and not-here, has remained on my mind. The body's precarious flux is always hovering between in-breath and out-breath, between birth and death—we could go at any time.

Elsewhere, I have argued that poetry is most essentially a forming of attention in the medium of language. What is sometimes lost in theorizations of attention, though, is the fact that attention takes place in the body and according to the changing reach and limits of the finite physical brain. As both imprint and extension of the physicality of attention, poetic language is deeply rooted, transhistorically and microhistorically, in the body's vital signs (rhythm links to pulse, line length to breath, for example). Poetry depends on consciousness, the ability of the body (and its brain) to form language and respond to stimuli (responsiveness). Perhaps, in tending and attending to an unresponsive or unconscious body, the lack or limitedness of language (either in the muteness of worried watching or in states of aphasia) makes the physicality of words more palpable.

Much has been written on the ancient custom of keeping vigil beside a beloved's body after life has passed from it or while the soul is lingering nearby. In the Judaic tradition, this vigil is accompanied by the speaking of words, the recitation of Tehillim—saying Kaddish—in rites of mourning that continue for eleven Hebrew months after the death of a parent or for thirty days after the death of a spouse, sibling, or child. The Psalms themselves are poem-songs of praise that candle through the pain of loss. Death is not the subject but the occasion and impetus for speaking or singing these poems: in praise, the soul is kept turned toward God even as the heart grieves by the departed. To mourn without despair (without going with the dead into death) requires a splitting of attention between the life of the soul and the death of the body, between celebration and lamentation.

Much has been written, too, on the *poetry* of mourning—its formal rites and conventions, its changing historical significance, its inconclusiveness, particularly in the modern era.[1] But my investigation in this chapter is not about how we as both social and linguistic animals mourn (or try to mourn) the dead; it is about how we attend to life—a life that hangs, always, in such precarious balance—from the vantage of our own finite bodies, our own cognitive limits. And from there, how the making of language gives form and expression to this act of embodied attention. To give form to an instance of life (which is one of the things poems do) is to give form to a life whose passage is always already underway, as any moment is underway, and to give imprint of the unstable conditions that surround and shape a body's passage, moment to moment.

Sitting with my mother during both long comas, I was aware of the absolute lack of language in and for the scene of waiting and tending: not only the lack of my mother's language but also the deep muteness of my own mind, the utter failure of my brain to summon words that encompassed or even gestured toward the material and fleshly reality (or unreality) of that intensive care vigil. Seeking language for this waiting by and waiting with the living body, I came up short.

There is not a transhistorical or even a recent historical archive of coma poems, it seems. But there are poems about tending to the body, and there are poems that encircle an absent beloved—and between the presence of the dead body and the absence of the living body, I found myself eking out a place for language of the precarious body, or at least encircling the site of such language's absence. Reading about the dead, and reading about the longed-for living, and sitting with the precarious (living but comatose) body, I discovered a kinship between care given to the body out of which life has already passed, care given in physical absence, and the care given to the body whose life is only waveringly present.

While poetry of mourning performs and requires an act of attentional recollection—calling or hauling the name and accompanying life of the departed back from the past, or from the shades, against forgetting, the kind of attentional poiesis I'm interested in here is closer to vigilance—waiting, standing by, for an unguaranteed and unforeseeable change.[2] In this case, waiting for the body to wake up, to show signs of life independently of life support. Vigilance is a mode of attention held open. Its object is a field (or, for some attentional theorists,[3] a sphere) of potential into which some object or event might enter, or not. Vigilance is tiring, whether it takes place in a monastery, on a night watch's shift, before a blinking dashboard of surveillant screens, or by a hospital cot. It takes energy, not only from the mind (as though the mind could be separated from its flesh) but also from the metabolic systems, the joints, the back and neck muscles, the ocular nerves, the adrenal system. It is the *embodied* precarity induced by states of vigilance—heightened, perhaps, when the state is a covigilance shared by many caregivers surrounding a life that's hanging in the machine-regulated suspense of life support. *At-tendre*, to wait. To keep vigil. Invoking Malebranche, both Paul Celan and Simone Weil wrote of attention as "the natural prayer of the soul." The poet Cecilia Vicuña ties these threads together: "Precarious is what is obtained by prayer. Uncertain, exposed to hazards, insecure. From the Latin *precarius*, from *precis*; prayer."[4]

Being by the body—

In the opening passage of Mary Cornish's poem "The Lane," published in her 2007 volume, *Red Studio*, the speaker positions herself and the reader in a scene of being alone with the body of her husband after he has died. There is no description of setting or even of the room in which the body lies. There is only the speaker and the body, alone together:

> Left alone with his dead body,
> I took off my husband's socks,
> put my face on his feet.
> Unbuttoned his shirt, pulled down
> his pants, stroked and kissed the legs, chest,
> penis. There was nothing I did not want
> to hold, although in death his body
> had let go, the way I'd heard it did
> if a man were hanged. There was nothing
> that did not smell human.
> That night, his eyes were taken, and his skin
> removed for other bodies.[5]

The actions of the speaker are deliberate: the removal of socks, shirt, and pants, the stroking and kissing of legs, chest, and penis. The naming of each article and each part renders the body plainly and concretely present in these lines as matter. There is no mistaking this body for the living beloved. At the same time, the speaker's desire to touch, her recollection that "there was nothing I did not want / to hold" suggests a way in which the body still acts on her, still holds the place of the beloved who is absent. In the foreignness of death, the corpse both is and is not her husband. There is still love for this body, and there is also the discovery of its having "let go," "the way I'd heard it did / if a man were hanged." The living consciousness that was the beloved is not present, yet the flesh itself is there, in hand, to be touched and smelled. Just as "there was nothing" the speaker did not want to hold, "there was nothing that did not smell human." The proliferation of negatives and the reiteration of "there was nothing" in these lines materializes the husband's not-there-ness. This ambivalent negative presence causes the body of the beloved to waver between humanness and something else: it is there, he is not: "There was nothing that did not smell human." At the end of this passage, we learn that the eyes and skin (organs of sight and touch) will be given to others. Sight and touch: the very senses that the speaker has bestowed on the body in her vigil and (in this poem) transplanted into language will be transplanted as organs into the body of another, to see and to touch the bodies of yet others.

I have never washed, touched, or kept vigil beside a corpse. I can only imagine that it is quite a different thing to wash a body, to tend to the flesh, when the body has already "let go" than it is to tend to the living flesh in the half-light of life. The body in coma appears lifeless, but the flesh is warm; the movements of breathing and pulse (even if aided by machines) are visible and palpable, and bodily processes such as bowel movements, urination, and the growth of hair and nails are ongoing signs of life. There are also, of course, many differences between washing the body of a lover or spouse and washing the body of a parent. Yet both kinds of tending to the body (the living and the dead) are acts of embodied tending to a body one has known intimately, a body rendered both familiar and unfamiliar by the absence of consciousness.

Throughout *Red Studio*, in recounting the limbo position of the speaker between desire and bereavement, Cornish's poems of mourning present the grief vigil as an attentional and physical state of suspension: being alongside an absence that is materially present as flesh and as remembered life; being both with the body and in a body, both with the dead and with the living. In this way, the grief vigil is also an act of intensive care—an emptying or suspension of the self in giving attention to another's body. Even as it tends to another's body, the body of the one keeping vigil inhabits its own precarious materiality, from the fragile vantage of its own physical and attentional faculties. In vigilance, the living body shares a suspended state with the sleeping or the dead.

In Cornish's poem "Tomb Painting," this embodied suspension of attention in mourning's extended vigil takes the form of the "Night Barge." The living wife is seated behind the dead husband, and they float together, living-dead, as if in sleep:

> The wife is not dead, yet she rides the Night Barge,
> chair behind her husband's chair,
> as if they still slept back-to-belly in their bed—
> that fit of curves and hollows we call *sleeping spoons*—
> the silvery way they moved, like fish,
> below the water of their sleep.[6]

Thinking about the embodied suspension of life in prolonged vigil-keeping, I have found myself reaching further backward into poetic history, into the lyric storytelling of Homeric epic and hymn. Perhaps because the space of myth (and even Homer himself, a figure both eternal and ephemeral, both historical and legendary) shares some of the qualities of here and not-here, real and unreal, for which I've been seeking language. I find myself turning and returning to the stories of two women who kept prolonged vigils, not by the dead but by a beloved's absence: Penelope and Demeter.

The weaving finespun / the yarns endless—

While Homer's *Odyssey* centers around Odysseus's decades-long struggle to return home, the epic is as much about the hero's absence, and the long years of awaiting his return, experienced by his father, son, and wife at home in Ithaca.[7] Indeed, the epic is already well underway before the protagonist makes his first appearance. It begins and circles back to those who wait at home, whose lives are shaped around this waiting. In the case of Penelope, this waiting is an embattled space—fending off suitors who have taken up residence in Odysseus's home, urging her to choose one of them and declare Odysseus dead. The suitors would like Penelope to move on, for time to move forward. Penelope guards against this, tending to Odysseus's absence and thereby holding time in suspension. To move on would be to foreclose the possibility of his return to the world, or at least to her world.

The performative strategy she devises for this temporal suspension can be seen as a formal materialization of the state of vigilance. To keep the suitors at bay, thereby buying herself and Odysseus time, Penelope begins to weave. She sets up a great loom and tells her suitors that she must complete the funeral shroud for Laertes before she can select one of them to wed. All day she weaves the fine cloth, and by night she unweaves her progress, so that the work cannot be completed.[8] This is the story in Penelope's words:

> I yearn for Odysseus, always, my heart pines away.
> They rush the marriage on, and I spin out my wiles.
> A god from the blue it was inspired me first
> to set up a great loom in our royal halls
> and I began to weave, and the weaving finespun,

> the yarns endless ...
>
> ...
>
> So by day I'd weave at my great and growing web—
> by night, by the light of torches set beside me,
> I would unravel all I'd done. Three whole years
> I deceived them blind, seduced them with this scheme.
> Then, when the wheeling seasons brought the fourth year on
> and the months waned and the long days came round once more,
> then, thanks to my maids—the shameless, reckless creatures—
> the suitors caught me in the act, denounced me harshly.
> So I finished it off. Against my will. They forced me.[9]

By the time Penelope tells this story, the reader has already heard it—almost word for word in the long stretches of exact repetition that make this epic song—from the mouth of her son Telemachus. When Penelope tells the story, she is speaking to Odysseus himself ("master of craft") in the guise of a stranger. She both is and is not addressing her long-awaited beloved, who is both there and not there.

Were it not punctured by one of her maids divulging Penelope's nocturnal unweaving to the suitors, her temporal suspension could have gone on forever, extending to the indefinite and unguaranteed possibility of Odysseus's return. This making-unmaking enables her to hold off time's passage in a sense—remaining by the side of her (living but absent) beloved in hopes of his return. Penelope's weaving and unweaving successfully fends off, for a time, the outside world's forward-moving impulse, holding open an enclosed space of waiting. In Fagles's translation, the forced halt of the vigilance-weaving is emphasized in the syntactical staccato of line 175, particularly in contrast to the extended sentence beginning with line 171 that stretches to encompass the "wheeling seasons" of four years.

Penelope's vigil is broken off and the shroud completed not because of the longed-for event of return but because of an external intrusion: the keeping of the vigil is out of step with and therefore intolerable to the forward thrust of the social and political realms. Penelope is caught in a double bind in which her stubborn devotion to her absent husband both proves her character's role as the perfect wife and makes her a hated-desired stick in the wheel of public social order and the timely transfer of property. The suspension or attenuation of time in attentional vigil-keeping jars with the surrounding temporalities and social weaves. Waiting *for* the absent beloved, or waiting *with* the beloved's absence, causes disorder and chaos to well up all around the site of waiting: the larders are emptied by a courtship party that won't end. The dishes and to-do lists pile up; the dissertation file languishes.

One can neither mourn nor cease mourning someone who has not (necessarily) died but who has also given no sign of life for a very long time. We can see in this kind of vigil a liminal attentional state stretched in suspension between grief and hope, without object or end. Penelope's vigil asks/wagers "is he/he is still out there," just as my family's ICU vigil asked/wagered "is she/she is still in there."

Cecilia Vicuña writes,

> The poverty of the thread
> was the limit
> and edge
> of the world
> was any moment.[10]

In vain much white barley—

In the story of Demeter and Persephone, we find an extended, cyclical vigil that shares with Penelope's the act of grieving a beloved who is not dead but whose absence is death-like. The story of abduction and partial (seasonal) return is well known: while picking flowers in a meadow with her friends, Demeter's daughter Persephone is seized by Hades and taken to his underworld domain to be his queen. The Homeric *Hymn to Demeter*[11] opens with an eleven-line sentence in which mother, daughter, flower, snare, and abductor are so syntactically dependent on one another (woven around the flower itself, the snare) that it is difficult to extract one from another:

> Demeter I begin to sing, the fair-tressed awesome goddess,
> herself and her slim-ankled daughter whom Aidoneus[12]
> seized; Zeus, heavy-thundering and mighty-voiced, gave her,
> without the consent of Demeter of the bright fruit and golden sword,
> as she played with the deep-breasted daughters of Ocean,
> plucking flowers in the lush meadow—roses, crocuses,
> and lovely violets, irises and hyacinths and the narcissus,
> which Earth grew as a snare for the flower-faced maiden
> in order to gratify by Zeus's design the Host-to-Many,
> a flower wondrous and bright, awesome for all to see,
> for the immortals above and for mortals below.
> From its root a hundredfold bloom sprang up and smelled
> so sweet that the whole vast heaven above
> and the whole earth laughed, and the salty swell of the sea.
> The girl marveled and stretched out both hands at once
> to take the lovely toy. The earth with its wide ways yawned

over the Nyssian plain; the lord Host-to-Many rose up on her
with his immortal horses, the celebrated son of Kronos;
he snatched the unwilling maid into his golden chariot
and led her off lamenting. She screamed with a shrill voice [. . .]
(lines 1–20)

The violence of the scene and its proximity to rape lend the hymn's insistence that Persephone's cries are heard by almost no one (mortal or immortal) yet more painful resonance.[13] Not even the "olives bright with fruit" hear her shrill screams. Hekate and Helios hear Persephone's cries, but they do not see what has happened. That her daughter's cries went unheard amplifies Demeter's maternal distress. Not only has the child been taken from maidenhood and from the realm of the living, but her mother does not hear her cries for help until they have echoed back from the mountainsides—by which point she is already gone:

> Sharp grief seized her heart, and she tore the veil
> on her ambrosial hair with her own hands.
> She cast a dark cloak on her shoulders
> and sped like a bird over dry land and sea,
> searching. No one was willing to tell her the truth,
> no one of the gods or mortals;
> no bird of omen came to her as truthful messenger.
> Then for nine days divine Deo[14] roamed over the earth,
> holding torches ablaze in her hands;
> in her grief she did not once taste ambrosia
> or nectar sweet-to-drink, nor bathed her skin. (lines 38–50)

It is here that Demeter's grief vigil begins. She roams the earth, veil torn from her hair, cloaked and bearing torches in her hands.

She cannot or will not eat, drink, or bathe. Her confused distress is bolstered by anger when she discovers the plight of her daughter, captive in the underworld. In its totalizing mixture of rage and bereavement, her grief vigil pushes the earth itself (Demeter's divine charge) into "a terrible and brutal year" of agricultural barrenness in which no seed will grow:

> . . . Then golden-haired Demeter
> remained sitting apart from all the immortals,
> wasting with desire for her deep-girt daughter.
> For mortals she ordained a terrible and brutal year
> on the deeply fertile earth. The ground released
> no seed, for bright-crowned Demeter kept it buried.
> In vain the oxen dragged many curved plows down
> the furrows. In vain much white barley fell on the earth. (lines 302–309)

In the end, Zeus comes through, and Hades returns Persephone to the living, but not before feeding her one pomegranate seed in the underworld.[15] The tasting of this single seed binds Persephone to his realm, so that, while she can ascend to her mother on Earth for two-thirds of each year, for the remaining portion she must return to Hades's kingdom. While Persephone is with the living, Demeter smiles her bounty on the earth: seeds sprout and crops flourish. When she descends again, Demeter's grief brings winter's barrenness to the land. Nothing will grow, no new life springs forth, while the mother attends to the absence of her daughter from the living world.

Demeter's vigil in Persephone's wavering, cyclical presence-absence in the world attends to both human frailty and environmental precarity—that what is given can be taken away, that the plenty of a beloved's presence and the warmth of their proximity can vanish so quickly. It thus links, too, the changing dynamics of attention (from recollecting and waiting for the absent to celebrating and adoring the return to presence) to environmental conditions. The recurrent season of lack wrought by Demeter's grief vigil each year suggests a co-implication of precarities. Her cycles of ascetic waiting lend the same cycles of scarcity and lack—the periods of waiting and wintering out. Her limbo position—having and then not having, being with and being without her beloved child—ripples outward to mortal life everywhere, dependent on Demeter's presence, dependent on Persephone's presence, to supply sustenance and growth. It is a webbed co-implication of precarities: the consequences of one precarity generating others. Yet the cyclical nature of loss and recovery generates a form in the wheel of the seasons. That two-thirds and one-third sum to a whole allows the myth to harmonize its vigil into something sustainable and sustaining. Demeter's vigil can go on forever, in this sense, because it does not go on forever.

In her collection, *Mother Love*, Rita Dove rethinks Persephone's abduction and Demeter's grief in the form of a series of sonnets, or variations/violations of the sonnet form.[16] Writing about her choice of the sonnet as the form for retelling this myth in "An Intact World," the preface to the volume, Dove writes, "The sonnet is a *heile Welt*, an intact world where everything is in sync, from the stars down to the tiniest mite on a blade of grass. And if the 'true' sonnet reflects the music of the spheres, it then follows that any variation from the strictly Petrarchan or Shakespearean forms represents a world gone awry." But Dove immediately questions this too-easy thinking about the inviolable stability of the sonnet: "Can't form also be a talisman against disintegration? The sonnet defends itself against the vicissitudes of fortune by its charmed structure, its beautiful bubble. All the while, though, chaos is lurking outside the gate." She suggests that urgent insistence on stability and everlastingness in the sonnet might be reread as an acknowledgment of the very instability and non-intactness of the world "outside the gate" of the verse form's fourteen lines. Linking this worldly chaos, against which the sonnet guards, to Demeter's grief, she writes that the ancient story suggests "a modern dilemma as well—there comes a point when a mother can no longer protect her child."[17] The same is later true in reverse: there comes a point when the daughter can no longer protect the mother from illness, from mortality. Dove writes, "The sonnet comforts even while its prim borders (but what a pretty fence!) are stultifying; one is constantly bumping up against Order. The Demeter/Persephone cycle of betrayal and regeneration is ideally suited for this form since all three—mother-goddess, daughter-consort and poet—are struggling to sing in their chains."[18]

> if I whispered to the moon
> I am waiting
> if I whispered to the olive
> you are on the way
> which would hear me?
>   I am listening
> the garden gone
> the seed in darkness
> the city around me
> I am waiting
>
>   . . .
>
> through whispers the sighing
> and let sorrow travel
> through sighing the darkness

be still she whispers
I am waiting
and light will enter
you are on your way (33)

While this chapter has focused on Demeter's attentive vigil, her long months of waiting, the passage from Dove's poem "Persephone in Hell" envisions a parallel vigil taking place in the underworld, as Persephone waits for her mother, waits to return to the living. At the bedside of the unconscious, we wait for the loved one to return to the living. At the bedside of the dead, we wait for the soul to pass from the body. In the coma's underworld, the one on life support undergoes a different vigil perhaps—one for which there is no language because no stable pathway of consciousness or memory exists between this world and that. We attend to what we can, while we can.

Five years have passed since my mother's comas.[19] Had she fallen ill this year, none of us would have been able to sit with her during those months of unconsciousness. There would have been no family bedside vigil, only the industrious, exhausted, inexhaustible work of health care workers getting through the COVID-19 pandemic. I wonder, sometimes, whether the absence of family members milling about the ICU floors makes the nurses' jobs harder or (more likely) easier. I also bear in mind the fact that hospice workers shouldered this labor vigil (the constant care they keep in the days, weeks, months, and years between a patient's active life and their death) for a long time before the pandemic and will continue to do afterward.

My mother has significant cognitive damage but is still able to live at home—and insists on doing so out of characteristic stubbornness, despite all our worries about what the next fall will bring. The walls of the room where she sleeps (first a small room upstairs that used to be my brother's bedroom, now a cot set up downstairs in the living room because she can no longer safely navigate stairs) are filled with vibrant pages cut out from the coloring books that fill her mornings and afternoons—intricate designs of flora and fauna: a snake with mosaic scales, a leaf bouquet with tiny, coiled tendrils, a ship on a curling wave, filled in with pencils, markers, and a patience I could not muster if I tried. Aqua vein, sienna leaf, silver sail, lime fin.

Describing the long-term hospice care of his partner, Wally, in *Still Life with Oysters and Lemon*,[20] Mark Doty writes about the way rooms change to accommodate illness—its long management, the rearrangements it renders choiceless in the lives of those who love and tend to loved ones:

> In a while our living room became a bedroom, since Wally could no longer walk up stairs. Then, our big four-poster had to go, since he needed a hospital bed we could crank up and down. I put a single iron bed next to it so we could sleep side by side, and did what I could to keep the room looking like home, but increasingly the sickroom stuff intruded, pushing other things away. . . .
>
> Later the room changed again—became my bedroom only, after Wally's death, the hospital bed dismantled and trucked away by the rental company, the room strangely resonant, vibrant . . . how else could I name the odd, profound sense of life in that room, life lived out and through? . . . Wally wasn't the first to die there, but I felt his death filled the space with a strange, vital light—a light inside the light—so that it became (I can say it no other way) vibrant. (42)

The colors in my mother's coloring books are (yes) *vibrant* (figures 14.1 and 14.2). Vibrant matter given vital form through the act of making. One afternoon she calls me from Virginia to say she has something important to say that doesn't make sense, but she thinks I might understand: "I am coloring all day long. They are just coloring books, but there is something about this coloring. It is like . . . a poem. When I'm coloring, it feels like a poem. Do you see?" Yes, I think I do.

FIGURE 14.1 A page from Lucy Alford's mother's coloring books

FIGURE 14.2 A few pages from Lucy Alford's mother's coloring books

NOTES

1. For a transhistorical study of the English elegiac tradition with special attention paid to inconsolability and irresolvability in the "work of mourning," see Peter Sacks, *The English Elegy: Studies in the Genre from Spenser to Yeats* (Baltimore, MD: Johns Hopkins University Press, 1985). Jahan Ramazani focuses on the specifically modern dimensions of elegy in *Poetry of Mourning: The Modern Elegy from Hardy to Heaney* (Chicago: University of Chicago Press, 1994); R. Clifton Spargo analyzes the social and psychological ethics of transhistorical literary grief in *The Ethics of Mourning: Grief and Responsibility in Elegiac Literature* (Baltimore, MD: Johns Hopkins University Press, 2004); and Sandra Gilbert has compiled a far-reaching anthology of elegies in English, with primary (though not exclusive) emphasis given to modern works: *Inventions of Farewell: A Book of Elegies* (New York: Norton, 2001). For a study of elegiac work as a mode of poetic attention, see chapter 4 of my own *Forms of Poetic Attention* (New York: Columbia University Press, 2020), 98–124.
2. The act of attending in and to precarious bodies in scenes of intensive care such as those involving life support, hospice, and chronic illness situates us among several modes of attending—notably vigilance, desire, and recollection—at once. Vigilance casts into the future, recollection into the past, and desire into absence. These temporal/attentional castings take place in a body marked by both potentiality and finitude, perched in the intersectional present moment that is itself always in the midst of coming into being and falling away. I offer a fuller discussion of vigilance as a mode of attention in chapter 7 of *Forms of Poetic Attention*.
3. Early in the emergence of the interdisciplinary field of attention studies, P. Sven Arvidson's *The Sphere of Attention: Context and Margins* (Dordrecht: Springer, 2006) bridged cognitive and phenomenological approaches to put forward a model of attention as sphere rather than field/spotlight or foreground/backdrop.
4. Cecilia Vicuña, "Entering," in *New and Selected Poems*, ed. and trans. Rosa Alcalá (Berkeley, CA: Kelsey Street Press, 2018), 85.
5. Mary Cornish, "The Lane," in *Red Studio* (Oberlin, OH: Oberlin College Press, 2007), 4.

6. Mary Cornish, "Tomb Painting," in *Red Studio* (Oberlin, OH: Oberlin College Press, 2007), 8.
7. Throughout this section, I refer to Robert Fagles's translation of *The Odyssey* (New York: Penguin, 1996).
8. Reading the lines "Generations of the imagination piled / In the manner of its stitchings, of its thread, / In the weaving round the wonder of its need" in Wallace Stevens's elegy for Henry Church, Peter Sacks has noted that "to speak of weaving a consolation recalls the actual weaving of burial clothes and shroud, and this emphasizes how mourning is an action, a process of work" (Sacks, *The English Elegy*, 19). For this reason, "each elegy is to be regarded . . . as a *work*, both in the commonly accepted meaning of a product and in . . . the sense that underlies Freud's phrase 'the work of mourning'" (Sacks, *The English Elegy*, 1, citing Sigmund Freud, "Mourning and Melancholia," in *General Psychological Theory*, trans. Joan Riviere, ed. Philip Rieff [1917; repr., New York: Collier, 1963], 164–79.)
9. Homer, *The Odyssey*, 19:167–175.
10. Cecilia Vicuña, from *quipoem*, in *The Precarious*, 36.
11. In this section, I draw on Helen Foley's lyrical translation of *The Homeric Hymn to Demeter* (Princeton, NJ: Princeton University Press, 1994), as well as her extensive commentary in the same volume. I consulted Martin West's prose translation of the *Homeric Hymns* (Cambridge, MA: Harvard University Press, 2003) in parallel with Foley's.
12. Hades.
13. Whether Persephone's virginity is *also* stolen by force in this violent scene, and at what point the nonconsensual marriage is physically consummated, are left unclear. In her commentary on the *Hymn to Demeter*, Helen Foley points out that, while nonconsensual marriage is not uncommon in Greek myth, the hymn dwells on the forcefulness of the theft, emphasizing by proximity the contrast between Zeus *giving* Persephone to be Hades's wife and Hades *seizing* her midplay, without her mother's knowledge (Foley, *The Homeric Hymn*, 31). Certainly, for contemporary readers, the double theft of childhood and life-among-the-living cannot help but summon up events of rape and the subsequent isolation of trauma's shadow world.
14. Demeter.

15. The language of this deceptive seed-feeding makes it difficult (for me, at least) to determine whether the seed is snuck secretly into Persephone's mouth and whether she knows she should not eat it (is she tempted by its sweetness and its smallness, or is she unaware that it is in her mouth until too late?). West translates the moment as "But he gave her a honey-sweet pomegranate seed to eat, surreptitiously, peering about him" (*Hymn to Demeter*, lines 371–373; West, 61). Foley renders it as "But he gave her to eat / a honey-sweet pomegranate seed, stealthily passing it / around her" (*Hymn to Demeter*, lines 371–373; Foley, 20).
16. Rita Dove, *Mother Love* (New York: Norton, 1995).
17. Dove, Mother Love, 1.
18. Dove, Mother Love, 2.
19. There is always a temporal gap between the moment of writing and the date of publication. In this case, the gap is also a passage. This essay was written in 2020. My mother died on January 20, 2023, while this volume was in production. Beyond this note, I have chosen not to alter the essay itself.
20. Mark Doty, *Still Life with Oysters and Lemon* (Boston: Beacon, 2001).

BIBLIOGRAPHY

Abel, Jacob Friedrich. *Eine Quellenedition zum Philosophieunterricht an der Stuttgarter Karlsschule (1773–1782)*, ed. Wolfgang Riedel. Würzburg: Königshausen & Neumann, 1995.

ACS-ED District Demographic Dashboard 2014–18. "Providence School District, RI." National Center for Education Statistics. Accessed December 7, 2020. https://nces.ed.gov/Programs/Edge/ACSDashboard/4400900.

Adams, Grace. "Titchener at Cornell." *American Mercury* 24, no. 96 (1931).

Adler, Hans. "Bändigung des (Un)Möglichen: Die Ambivalente Beziehung zwischen Aufmerksamkeit und Aufklärung." In *Reiz, Imagination, Aufmerksamkeit: Erregung und Steuerung von Einbildungskraft im klassichen Zeitalter (1680–1830)*, ed. Jörn Steigerwald and Daniela Watzke, 41–54. Würzburg: Königshausen & Neumann, 2003.

Ahmed, Sara. "A Phenomenology of Whiteness." *Feminist Theory* 8, no. 2 (2007): 149–68. https://doi.org/10.1177/1464700107078139.

Alford, Lucy. *Forms of Poetic Attention*. New York: Columbia University Press, 2020.

American Psychiatric Association, *Diagnostic and Statistical Manual of Mental Disorders*, 3rd ed. Washington, DC: American Psychiatric Association, 1980.

———. *Diagnostic and Statistical Manual of Mental Disorders*, 4th ed. Washington, DC: American Psychiatric Association, 1994.

———. *Diagnostic and Statistical Manual of Mental Disorders*, 5th ed. Washington, DC: American Psychiatric Association, 2012.

Andelfinger, John A., Masha Berman, Larissa A. Bray, Nathan E. Kruis, Mallory O. Smith, and Paul M. Hawkins. *Perceptions of the Summit Learning Platform*. Indiana: Indiana University of Pennsylvania, 2018.

Angell, James Rowland. "Habit and Attention." *Psychological Review* 5, no. 2 (1898): 179–83.

Angell, James Rowland, Addison W. Moore. "Reaction-Time: A Study in Attention and Habit." *Psychological Review* 3, no. 3 (1896): 251.

Arpaly, Nomy. *Unprincipled Virtue: An Inquiry into Moral Agency*. New York: Oxford University Press, 2002.

Arvidson, P. Sven. "Experimental Evidence for Three Dimensions of Attention." In *Gurwitsch's Relevancy for Cognitive Science*, ed. Lester Embree, 151–68. Dordrecht: Springer, 2004.

———. *The Sphere of Attention: Context and Margin*. Dordrecht: Springer, 2006.

Backe, Andrew. "John Dewey and Early Chicago Functionalism." *History of Psychology* 4, no. 4 (2001): 323–40.

Badre, David. "Tips from Neuroscience to Keep You Focused on Hard Tasks." *Nature*, March 15, 2021. https://www.nature.com/articles/d41586-021-00606-x.

Bahdanau, Dzmitry, Kyunghyun Cho, and Yoshua Bengio. "Neural Machine Translation by Jointly Learning to Align and Translate." ArXiv:1409.0473 [Cs, Stat]. May 19, 2016: 4. https://doi.org/10.48550/arXiv.1409.0473.

Baldwin, J. Mark. "Types of Reaction." *Psychological Review* 2, no. 3 (1895): 259–73.

Ballentine, Kess L. "Understanding Racial Differences in Diagnosing ODD Versus ADHD Using Critical Race Theory." *Families in Society: The Journal of Contemporary Social Services* 100, no. 3 (2019): 282–92.

Barber, Elizabeth Wayland. *Women's Work: The First 20,000 Years—Women, Cloth, and Society in Early Times*. New York: Norton, 1995.

Barrow Jr., Mark V. *A Passion for Birds: American Ornithology After Audubon*. Princeton, NJ: Princeton University Press, 1998.

Barthes, Roland. "The Grain of the Voice." In *Image, Music, Text*. trans. Stephen Heath, 179–89. New York: Hill and Wang, 1977.

Bauchner, Joshua. "Fechner on a Walk: Everyday Investigations of the Mind-Body Relationship." *Historical Studies in the Natural Sciences* 51, no. 1 (2021): 1–47.

Baxandall, Michael. *The Limewood Sculptors of Renaissance Germany*. New Haven, CT: Yale University Press, 1980.

Beaty, Roger E., et al. "Robust Prediction of Individual Creative Ability from Brain Functional Connectivity." *Proceedings of the National Academy of Sciences* 115, no. 5 (2018): 1087–92.

Beiser, Frederick. "Herbart's Monadology." *British Journal for the History of Philosophy* 23 (2015): 1056–73.

Benschop, Ruth, and Douwe Draaisma. "In Pursuit of Precision: The Calibration of Minds and Machines in Late Nineteenth-Century Psychology." *Annals of Science* 57 (2000): 1–25.

Benson, Etienne. "A Centrifuge of Calculation: Managing Data and Enthusiasm in Early Twentieth-Century Bird Banding." *Osiris* 32, no. 1 (2017): 286–306.

Berdik, Chris. "Tipping Point: Can Summit Put Personalized Learning Over the Top?" *Hechinger Report*. January 17, 2017. https://hechingerreport.org/tipping-point-can-summit-put-personalized-learning-top/.

Berlant, Lauren. *Cruel Optimism*. Durham, NC: Duke University Press, 2011.

Bermúdez, Juan Pablo. "Social Media and Self-Control: The Vices and Virtues of Attention." In *Social Media and Your Brain: Web-Based Communication Is Changing How We Think and Express Ourselves*, ed. C. G. Prado, 57–74. Santa Barbara, CA: Praeger, 2017.

Berwick, Robert C., and Noam Chomsky. *Why Only Us: Language and Evolution*. Cambridge, MA: MIT Press, 2016.

Bhattacharyya, K. C. "Some Aspects of Negation." In *Studies in Philosophy*, vol. 2, ed. Gopinath Bhattacharyya, 205–20. Calcutta: Progressive, 1958. First published in 1914.

———. "The Subject as Freedom." In *Studies in Philosophy*, vol. 2, ed. Gopinath Bhattacharyya, 19–94. Calcutta: Progressive, 1958. First published in 1930.

Bhikkhu, Thanissaro, trans. *Samaññaphala Sutta: The Fruits of the Contemplative Life*, DN 2 PTS D i 47, Access to Insight, 1997. https://www.accesstoinsight.org/tipitaka/dn/dn.02.0.than.html.

———. *The Wings to Awakening: An Anthology from the Pali Canon*. Valley Center, CA: Metta Forest Monastery, 1996.

Bill & Melinda Gates Foundation. *A Working Definition of Personalized Learning*. Verona, NJ: New Classrooms, 2018. https://www.newclassrooms.org/wp-content/uploads/2018/08/personalized-learning-working-definition-1.pdf.

Bjerre-Nielsen, Andreas, et al. "The Negative Effect of Smartphone Use on Academic Performance May Be Overestimated: Evidence from a 2-Year Panel Study." *Psychological Science* 31, no. 11 (2020): 1351–62.

Blatter, Jeremy T. "The Psychotechnics of Everyday Life: Hugo Münsterberg and the Politics of Applied Psychology, 1887–1917." PhD diss., Harvard University, 2014.

Bordogna, Francesca. *William James at the Boundaries: Philosophy, Science, and the Geography of Knowledge*. Chicago: University of Chicago Press, 2008.

Boring, Edwin G. *A History of Experimental Psychology*. New York: Century, 1929.

———. *Psychologist at Large*. New York: Basic Books, 1961.

Brahm, Ajahn. *Mindfulness, Bliss, and Beyond: A Meditator's Handbook*. New York: Simon & Schuster, 2006.

Brasington, Leigh. *Right Concentration: A Practical Guide to the Jhānas*. Boulder, CO: Shambhala, 2015.

Bratich, Jack Z. *Conspiracy Panics: Political Rationality and Popular Culture*. Albany, NY: SUNY Press, 2008.

Braun, Erik. *The Birth of Insight: Meditation, Modern Buddhism, and the Burmese Monk Ledi Sayadaw*. Chicago: University of Chicago Press, 2013.

———. "The United States of Jhāna." In *Buddhism Beyond Borders: New Perspectives on Buddhism in the United States*, ed. Scott Mitchell and Natalie Quli. 163–80. Albany: SUNY Press, 2015.

Braunschweiger, David. *Die Lehre von Aufmerksamkeit in der Psychologie des 18. Jahrhunderts*. Leipzig: Hermann Haacke, 1899.

Bredo, Eric. "Evolution, Psychology, and John Dewey's Critique of the Reflex Arc Concept." *Elementary School Journal* 98, no. 5 (1998): 447–66.

Brenninkmeijer, Jonna. *Neurotechnologies of the Self: Mind, Brain and Subjectivity*. London: Palgrave Macmillan, 2016.

Brock, André, Jr. *Distributed Blackness: African American Cybercultures*. New York: NYU Press, 2020.

Brown, Judith K. "A Note on the Division of Labor by Sex." *American Anthropologist* 27, no. 5 (1970): 1073–78.

Bruya, Brian. *Effortless Attention: A New Perspective in the Cognitive Science of Attention and Action*. Cambridge, MA: MIT Press, 2010.

Bruyninckx, Joeri. *Listening in the Field: Recording and the Science of Birdsong*. Cambridge, MA: MIT Press, 2018.

Burbea, Rob. "Retreat Dharma Talks: Practising the Jhānas." *Dharma Seed*, December 2019. https://dharmaseed.org/retreats/4496/ (recordings of dharma talks from a twenty-three-day jhāna retreat at Gaia House in Devon, UK).

———. *Seeing That Frees: Meditations on Emptiness and Dependent Arising.* Devon: Hermes Amāra, 2014.

Burkeman, Oliver. "Commercial Interests Exploit a Limited Resource on an Industrial Scale: Your Attention." *Guardian,* April 1, 2015. https://www.theguardian.com/commentisfree/oliver-burkeman-column/2015/apr/01/commercial-interests-exploit-limited-resource-attention.

Burnett, D. Graham. "Induced Attention." In *Nora Turato: Explained Away. Pool 3,* ed. Fabian Flueckiger, 1–33. Liechtenstein: Kunstmuseum Liechtenstein, 2019.

Burnett, D. Graham, and Stevie Knauss, eds. *Twelve Theses on Attention.* Princeton, NJ: Princeton University Press, 2022.

Burnham, John C. "Thorndike's Puzzle Boxes." *Journal of the History of the Behavioral Sciences* 8, no. 2 (1972): 159–67.

Cain, Jeff. "It's Time to Confront Student Mental Health Issues Associated with Smartphones and Social Media." *American Journal of Pharmaceutical Education* 82, no. 7 (2018): 6862.

Catherine, Shaila. *Focused and Fearless: A Meditator's Guide to States of Deep Joy, Calm, and Clarity.* New York: Simon & Schuster, 2008.

Cerullo, John J. "E. G. Boring: Reflections on a Discipline Builder." *American Journal of Psychology* 101 (1988): 563.

Chan, Anita Say. "Venture Ed: Recycling Hype, Fixing Futures, and the Temporal Order of Edtech." In *digitalSTS: A Field Guide for Science and Technology Studies,* ed. Janet Vertesi and David Ribes, 161–77. Princeton, NJ: Princeton University Press, 2019.

Chin, Kate. quoted in The Learning Network, "What Students Are Saying About: Online Learning, Family Vacations and Moving to a New Home." *New York Times,* May 2, 2019. https://www.nytimes.com/2019/05/02/learning/what-students-are-saying-about-online-learning-family-vacations-and-moving-to-a-new-home.html.

Chomsky, Noam. "Endorsement by La Mettrie." In *Man a Machine and Man a Plant.* trans. Richard Watson and Maya Rybalka. Indianapolis: Hackett, 1994.

Christoff, Kalina, et al. "Experience Sampling During fMRI Reveals Default Network and Executive System Contributions to Mind Wandering." *Proceedings of the National Academy of Sciences* 106, no. 21 (2009): 8719–24.

Chrysikou, Evangelia G. "The Costs and Benefits of Cognitive Control for Creativity." In *The Cambridge Handbook of the Neuroscience of Creativity,* ed. Rex E. Jung and Oshin Vartanian, 195–210. Cambridge: Cambridge University Press, 2018.

Citton, Yves. *The Ecology of Attention*. trans. Barnaby Norman. Cambridge: Polity, 2017. First published in French as *Pour une écologie de l'attention* in 2014 by Éditions du Seuil (Paris).

Clark, Herbert H. *Using Language*. Cambridge: Cambridge University Press, 1996.

Coen, Deborah R. *Vienna in the Age of Uncertainty: Science, Liberalism, and Private Life*. Chicago: University of Chicago Press, 2007.

Cooke, Edward S., Jr. "Artisanal Time: Cumulative, Partially Invisible, Non-linear, and Episodic." In *Marking Time: Objects, People, and Their Lives, 1500–1800*, ed. Edward Town and Angela McShane, 83. New Haven, CT: Yale University Press, 2020.

Coon, Deborah J. "Standardizing the Subject: Experimental Psychologists, Introspection, and the Quest for a Technoscientific Ideal." *Technology and Culture* 34, no. 4 (1993): 757–83.

Cornish, Mary. *Red Studio*. Oberlin, OH: Oberlin College Press, 2007.

——. "The Lane." In *Red Studio*, 4. Oberlin, OH: Oberlin College Press, 2007.

——. "Tomb Painting." In *Red Studio*, 8. Oberlin, OH: Oberlin College Press, 2007.

Crary, Jonathan. "Spectacle, Attention, Counter-Memory." *October* 50 (1989): 97–107.

——. *Suspensions of Perception: Attention, Spectacle, and Modern Culture*. Cambridge, MA: MIT Press, 1999.

——. *Suspensions of Perception: Attention, Spectacle, and Modern Culture*. Cambridge, MA: MIT Press, 2001.

——. *Techniques of the Observer: On Vision and Modernity in the Nineteenth Century*. Cambridge, MA: MIT Press, 1990.

Csikszentmihalyi, Mihaly. *Flow: The Psychology of Optimal Experience*. New York: Harper & Row, 1990.

Dahlstrom, Daniel O., ed. *Moses Mendelssohn: Philosophical Writings*. Cambridge: Cambridge University Press, 1997.

Danckert, James, and John D. Eastwood, *Out of My Skull: The Psychology of Boredom*. Cambridge, MA: Harvard University Press, 2020.

Danziger, Kurt. *Constructing the Subject: Historical Origins of Psychological Research*. Cambridge: Cambridge University Press, 1990.

Darwin, Charles. *The Descent of Man, and Selection in Relation to Sex.* 1871; repr. Princeton, NJ: Princeton University Press, 1981.

Daston, Lorraine, and Peter Galison. *Objectivity.* New York: Zone, 2007.

De La Mettrie, Julien Offray. *Man a Machine and Man a Plant.* trans. Richard A. Watson and Maya Rybalka. Indianapolis, IN: Hackett, 1994. Both works first published in 1748.

Deleuze, Gilles, and Félix Guattari. *A Thousand Plateaus: Capitalism and Schizophrenia.* trans. Brian Massumi. London: Continuum, 2004.

Dessoir, Max. *Geschichte der neueren deutschen Psychologie,* 2nd ed. 1894; repr. Berlin: Carl Duncker, 1902.

De Vignemont, Frederique, ed. *The World at Our Fingertips: A Multidisciplinary Exploration of Peripersonal Space.* Oxford: Oxford University Press, 2021.

De Waal, Frans. *Mama's Last Hug: Animal Emotions and What They Tell Us About Ourselves.* New York: Norton, 2019.

Dewey, John. "The Reflex Arc Concept in Psychology." *Psychological Review* 3, no. 4 (July 1896): 369.

Dewhurst, Kenneth, and Nigel Reeves. *Friedrich Schiller: Medicine, Psychology, and Literature.* Berkeley: University of California Press, 1978.

Desjarlais, Robert. "Struggling Along: The Possibilities for Experience Among the Homeless Mentally Ill." *American Anthropologist* 96, no. 4 (1994): 886–901.

Desjarlais, Robert, and Jason Throop. "Phenomenological Approaches in Anthropology." *Annual Review of Anthropology* 40 (2011): 87–102.

Dilthey, Wilhelm. "Freidrich der Große und die deutsche Aufklärung." In *Wilhem Dilthey Gesammelte Schriften,* vol. 3. Stuttgart: Teubner, 1992 (1927).

Dobrynin, N. F. "Basic Problems in the Psychology of Attention." In *Psychological Science in the USSR,* vol. 1, 274–91. ed. B. G. Anan'yev. Washington, DC: U.S. Department of Commerce, Clearinghouse for Federal Scientific and Technical Information, 1966.

Doty, Mark. *Still Life with Oysters and Lemon.* Boston: Beacon, 2001.

Dove, Rita. *Mother Love.* New York: Norton, 1995.

Doyle, Jennifer, and David J. Getsy. "Queer Formalisms: Jennifer Doyle and David Getsy in Conversation." *Art Journal* 72, no. 4 (2013): 58–71.

Dreyfus, Hubert. "The Myth of the Pervasiveness of the Mental." In *Mind, Reason, and Being-in-the World: The McDowell-Dreyfus Debate,* ed. Joseph K. Schear. London: Routledge, 2013.

Dunlap, Thomas R. *In the Field, Among the Feathered: A History of Birders and Their Guides.* Oxford: Oxford University Press, 2011.

DuPaul, George J. "Adult Ratings of Child ADHD Symptoms: Importance of Race, Role, and Context." *Journal of Abnormal Child Psychology* 48, no. 5 (2020): 673–77.

DuPaul, George J., Qiong Fu, Arthur D. Anastopoulos, Robert Reid, and Thomas J. Power. "ADHD Parent and Teacher Symptom Ratings: Differential Item Functioning Across Gender, Age, Race, and Ethnicity." *Journal of Abnormal Child Psychology* 48, no. 5 (2020): 679–91.

Erdheim, Mario. "On the Problem of Free-Floating Attention." *Psyche* 42, no. 3 (1988): 221–24.

Erdmann, Johann. *A History of Philosophy*, 3rd English ed., vol. 2. trans. Williston S. Hough. London: Swan Sonnenschein, 1892 (1890).

Eyal, Nir, and Julie Li. *Indistractable: How to Control Your Attention and Choose Your Life.* London: Bloomsbury, 2020.

Fairhurst, Merle T., et al. "Confidence Is Higher in Touch Than in Vision in Cases of Perceptual Ambiguity." *Scientific Reports* 8 (2018): 15604.

Fairweather, Abrol, and Carlos Montemayor. *Knowledge, Dexterity, and Attention: A Theory of Epistemic Agency.* Cambridge: Cambridge University Press, 2017.

Fanon, Frantz. *Black Skin, White Masks.* London: Pluto, 2008.

Farrell, Maria. "The Prodigal Techbro." *The Conversationalist* (blog), March 5, 2020. https://conversationalist.org/2020/03/05/the-prodigal-techbro/.

FeelZing. "FeelZing Patch—Smart Energy When You Need It." accessed January 19, 2021. https://feelzing.com/pages/science.

———. "How to Use the Patch." accessed January 19, 2021. https://feelzing.com/pages/patch.

FeelZing Neurostimulation Blog. "Electricity—The Brain and Body's Natural Language." https://feelzing.com/blogs/news/electricity-the-brain-and-body-s-natural-language.

Feder, Johann G. H. *Logik und Metaphysik.* Vienna: Johann Thomas Edl. Von Trattern, 1783.

Fernandez-Duque, Diego, and Mark L. Johnson. "Attention Metaphors: How Metaphors Guide the Cognitive Psychology of Attention." *Cognitive Science* 23, no. 1 (1999): 83–116.

———. "Cause and Effect Theories of Attention: The Role of Conceptual Metaphors." *Review of General Psychology* 6, no. 2 (2002): 152–65.

Figdor, Carrie. *Pieces of Mind: The Proper Domain of Psychological Predicates.* Oxford: Oxford University Press, 2018.

Foley, Helen, trans. and ed. *The Homeric Hymn to Demeter: Translation, Commentary, and Interpretive Essays.* Princeton, NJ: Princeton University Press, 1994.

Folland, Tom. "Robert Rauschenberg's Queer Modernism: The Early Combines and Decoration." *Art Bulletin* 92, no. 4 (2010): 348–65.

Foucault, Michel. "Technologies of the Self." In *Technologies of the Self: A Seminar with Michel Foucault,* ed. Luther H. Martin, Huck Gutman, and Patrick H. Hutton. Amherst: University of Massachusetts Press.

Fowler, Geoffrey A. "This Gadget Gives You a Low-Voltage Pick-Me-Up." *Wall Street Journal,* July 21, 2015. https://www.wsj.com/articles/this-gadget-gives-you-a-low-voltage-pick-me-up-1437503825.

Freud, Sigmund. *The Ego and the Id.* New York: Norton, 1989.

———. *The Interpretation of Dreams.* trans. and ed. James Strachey. 1955. First published in German as *Die Traumdeutung* in 1900 by Franz Deuticke (Vienna).

———. *The Letters of Sigmund Freud.* trans. Tania Stern and James Stern, ed. Ernst L. Freud. New York: Basic Books, 1960.

———. "Mourning and Melancholia." In *General Psychological Theory.* trans. Joan Riviere, ed. Philip Rieff, 164–79. New York: Collier, 1963. First published in 1917.

———. *New Introductory Lectures on Psychoanalysis,* vol. 15. trans. and ed. James Strachey, 90. London: Hogarth, 1963.

———. "Project for a Scientific Psychology." In *Pre-Psycho-Analytic Publications and Unpublished Drafts (1886–1899).* Vol. 1 of *The Complete Psychological Works of Sigmund Freud.* 1950.

———. *Recommendations to Physicians Practising Psycho-analysis.* London: Hogarth, 1912.

———. *Two Encyclopaedia Articles.* London: Hogarth, 1955.

———. *The "Wolfman" and Other Cases.* trans. Louise Adey Huish. London: Penguin, 2003.

Fried, Michael. *Absorption and Theatricality: Painting and the Beholder in the Age of Diderot.* Berkeley: University of California Press, 1980.

———. "Art and Objecthood (1967)." In *Art and Objecthood,* 166–67. Chicago: University of Chicago Press, 1998.

Friends of Attention. "Twelve Theses on Attention." August 2019. https://friendsofattention.net/sites/default/files/2020-05/TWELVE-THESES-ON-ATTENTION-2019.pdf.

———. *Twelve Theses on Attention*, ed. D. Graham Burnett and Stevie Knauss. Princeton, NJ: Princeton University Press, 2022.

Fryxell, Allegra R. P. "Psychopathologies of Time: Defining Mental Illness in Early 20th-Century Psychiatry." *History of the Human Sciences* 32, no. 2 (2019): 3–31.

Gable, Shelly L., Elizabeth A. Hopper, and Jonathan W. Schooler. "When the Muses Strike: Creative Ideas of Physicists and Writers Routinely Occur During Mind Wandering." *Psychological Science* 30, no. 3 (2019): 396–404.

Gale, Matthew, and Chris Stephens, eds. *Barbara Hepworth: Works in the Tate Gallery Collection and the Barbara Hepworth Museum, St. Ives*. London: Tate Gallery, 1999.

Galison, Peter. "The Ontology of the Enemy: Norbert Wiener and the Cybernetic Vision." *Critical Inquiry* 21, no. 1 (1994): 228–66.

Gallistel, Charles R. *The Organization of Learning*. Cambridge, MA: MIT Press, 1990.

Ganeri, Jonardon. *Attention, Not Self*. Oxford: Oxford University Press, 2017.

Ganesh, Maya Indira. "The Center for Humane Technology Doesn't Want Your Attention."

Cyborgology (blog), February 9, 2018. https://thesocietypages.org/cyborgology/2018/02/09/the-center-for-humane-technology-doesnt-want-your-attention/.

Garfield, Monica J., et al. "Research Report: Modifying Paradigms—Individual Differences, Creativity Techniques, and Exposure to Ideas in Group Idea Generation." *Information Systems Research* 12, no. 3 (2001): 322–33.

Garve, Christian. "Ueber die Geduld." In *Christian Garves Sämmtliche Werke*, 1:45–46, 85–86. Breslau, 1801.

———. "Ueber die Muße." In *Anthologie aus den sämmtlichen Werken von Christian Garve*. Hildburghausen: 1844.

Gasché, Rodolphe. "One Seeing Away: Attention and Abstraction in Kant." *Centennial Review* 8 (2008): 1–28.

Gaut, Berys. "The Philosophy of Creativity." *Philosophy Compass* 5 (2010): 1034–46.

Getsy, David J. *Abstract Bodies: Sixties Sculpture in the Expanded Field of Gender.* New Haven, CT: Yale University Press, 2015.
Gilbert, Sandra. *Inventions of Farewell: A Book of Elegies.* New York: Norton, 2001.
Ginzburg, Carlo. "Clues: Roots of an Evidential Paradigm." In *Clues, Myths, and the Historical Method.* Baltimore, MD: Johns Hopkins University Press, 2013.
———. "Freud, the Wolf-Man, and the Werewolves." In *Clues, Myths, and the Historical Method.* Baltimore, MD: Johns Hopkins University Press, 2013.
Gleig, Ann. *American Dharma: Buddhism Beyond Modernity.* New Haven, CT: Yale University Press, 2019.
———. "From Buddhist Hippies to Buddhist Geeks: The Emergence of Buddhist Postmodernism?," *Journal of Global Buddhism* 15 (2014): 22.
Goodstein, Elizabeth S. *Experience without Qualities: Boredom and Modernity.* Stanford, CA: Stanford University Press, 2005.
Green, Christopher D. "Scientific Objectivity and E. B. Titchener's Experimental Psychology." *Isis* 101, no. 4 (2010): 697–721.
Guenther, Katja. *Localization and Its Discontents: A Genealogy of Psychoanalysis and the Neuro Disciplines.* Chicago: University of Chicago Press, 2015.
Gurwitsch, Aron. "The Field of Consciousness." In *The Collected Works of Aron Gurwitsch*, ed. Richard Zaner and Lester Embree, 3:1–412. Dordrecht: Springer, 2010. First published in 1964 by Duquesne University Press (Pittsburgh, PA).
Hagerty, Michael R., et al. "Case Study of Ecstatic Meditation: fMRI and EEG Evidence of Self-Stimulating a Reward System." *Neural Plasticity* (2013). https://doi.org/10.1155/2013/653572.
Hagner, Michael. "Towards a History of Attention in Culture and Science." *MLN* 118 (2003): 670–87.
Hall, Gordon. "Object Lessons: Thinking Gender Variance through Minimalist Sculpture." *Art Journal* 72, no. 4 (2013): 46–57.
Harris, Cheryl. "Whiteness as Property." *Harvard Law Review* 106, no. 8 (1993): 1707–91.
Hatfield, Gary. "Attention in Early Scientific Psychology." *IRCS Technical Reports Series* 144 (1995): 1–35.
Herder, Johann Gottfried von. *Philosophical Writings.* trans. and ed. Michael N. Forster. Cambridge: Cambridge University Press, 2002.

Hesselberth, Pepita. "Detox." In *Uncertain Archives: Critical Keywords for Big Data*, ed. Nanna Bonde Thylstrup, Daniela Agostinho, Annie Ring, Catherine D'Ignazio, and Kristin Veel, 141–50. Cambridge, MA: MIT Press, 2021.

Heuman, Linda. "Don't Believe the Hype." *Tricycle*, October 1, 2014.

Hicks, Lucy. "Watch Amoebas Solve a Microscopic Version of London's Hampton Court Maze." *Science*, August 27, 2020. https://doi.org/10.1126/science.abe5316.

Holly, Michael Ann. "Notes from the Field: Materiality." *Art Bulletin* 95, no. 1 (2013): 15.

Homer. *The Odyssey*. trans. Robert Fagles. New York: Penguin, 1996.

Howard, Philip N. *Lie Machines: How to Save Democracy from Troll Armies, Deceitful Robots, Junk News Operations, and Political Operatives*. New Haven, CT: Yale University Press, 2020.

Hui, Alexandra. "From 'Wuh Wuh' to 'Hoo-Hoo' and the Rituals of Representing Bird Song, 1885–1925." In *Objects and Standards: On the Limitations and Effects of Fixing and Measuring Life*, ed. Tord Larsen et al., 231–54. Durham, NC: Carolina Academic Press, 2021.

———. "Listening to Extinction: Early Conservation Radio Sounds and the Silences of Species." *American Historical Review* 126, no. 4 (2021): 1371–95.

Hunter, Ian. *Rival Enlightenments: Civil and Metaphysical Philosophy in Early Modern Germany*. Cambridge: Cambridge University Press, 2004.

Hutchins, Amey. "Margaret Naumburg Papers Finding Aid." 2000. Kislak Center for Special Collections, Rare Books and Manuscripts, University of Pennsylvania.

Hwang, Tim. *Subprime Attention Crisis: Advertising and the Time Bomb at the Heart of the Internet*. New York: FSG Originals, 2020.

Hylan, John Perham. "The Fluctuation of Attention." Series of Monograph Supplements. *Psychological Review* 2, no. 2 (1898): 2.

Ingold, Tim. "The Textility of Making." *Cambridge Journal of Economics* 34 (2010): 92.

———. "Toward an Ecology of Materials." *Annual Review of Anthropology* 41 (2012): 433.

Ingram, Daniel. "The Dharma Overground." https://www.dharmaoverground.org/dharma-wiki/-/wiki/Main/arahat.

InteraXon Inc. "Anchors and Attentional Loops." module of online course "The Power of Attention." accessed February 12, 2021. ChooseMuse.com, 2019. https://choosemuse.com/the-power-of-attention/.
———. "Understanding Mental Fitness and Fatigue." module of online course "The Power of Attention." accessed February 12, 2021. ChooseMuse.com, 2019. https://choosemuse.com/the-power-of-attention/.
Irani, Lilly, and Rumman Chowdhury. "To Really 'Disrupt,' Tech Needs to Listen to Actual Researchers." *Wired*, June 26, 2019. https://www.wired.com/story/tech-needs-to-listen-to-actual-researchers/.
Iversen, Margaret. *Alois Riegl: Art History and Theory*. Cambridge, MA: MIT Press, 1993.
Jablonsky, Rebecca. "Mindbending: An Ethnography of Meditation Apps in an Age of Digital Distraction." PhD diss., Rensselaer Polytechnic Institute, 2020.
Jacobs, Nancy J. *Birders of Africa: History of a Network*. New Haven, CT: Yale University Press, 2016.
James, William. "Attention." In *The Principles of Psychology*. 1981.
———. "Great Men, Great Thoughts and the Environment." *Atlantic Monthly* 46, no. 276 (1880): 441–59.
———. *The Principles of Psychology*. New York: Henry Holt, 1890.
———. *The Principles of Psychology*, vol. 1. New York: Henry Holt, 1918.
———. *Talks to Teachers on Psychology and to Students on Some of Life's Ideals*. Cambridge, MA: Harvard University Press, 1983. First published in 1899 by Henry Holt (New York).
Jay, Martin. *Songs of Experience: Modern American and European Variations on a Universal Theme*. Berkeley: University of California Press, 2005.
Jennings, Carolyn Dicey. "Attention and Perceptual Organization." *Philosophical Studies: An International Journal for Philosophy in the Analytic Tradition* 172, no. 5 (2015): 1265–78.
Jennings, Carolyn Dicey. *The Attending Mind*. Cambridge: Cambridge University Press, 2020.
Johns Hopkins Institute for Education Policy. *Providence Public School District: A Review—June 2019*. Baltimore, MD: Johns Hopkins Institute for Education Policy, June 25, 2019. https://jscholarship.library.jhu.edu/bitstream/handle/1774.2/62961/ppsd-revised-final.pdf.

Johnson, Addie, and Robert W. Proctor. *Attention: Theory and Practice*. Thousand Oaks, CA: Sage, 2004.

Jonçich, Geraldine M. *The Sane Positivist: A Biography of Edward L. Thorndike*. Middletown, CT: Wesleyan University Press, 1968.

Jones, Amelia. "Art History/Art Criticism: Performing Meaning." In *Performing the Body/Performing the Text*, ed. Amelia Jones and Andrew Stephenson, 39–55. London: Routledge, 1999.

Kahneman, Daniel, Anne Treisman, and Brian J. Gibbs. "The Reviewing of Object Files: Object-Specific Integration of Information." *Cognitive Psychology* 24, no. 2 (1992): 175–219.

Kant, Immanuel. *Anthropology from a Pragmatic Point of View*. trans. and ed. Robert B. Louden. Cambridge: Cambridge University Press, 2006.

Kleinmintz, Oded M., Tal Ivancovsky, and Simone G. Shamay-Tsoory. "The Two-Fold Model of Creativity: The Neural Underpinnings of the Generation and Evaluation of Creative Ideas." *Current Opinion in Behavioral Sciences* 27 (2019): 131–38.

Koehler, Margaret. *Poetry of Attention in the Eighteenth Century*. New York: Palgrave Macmillan, 2012.

Kohn, Harry E. *Zur Theorie der Aufmerksamkeit*. Halle: Ehrhardt Karras, 1894.

Košenina, Alexander. *Ernst Platners Anthropologie und Philosophie: der philosophische Arzt und seine Wirkung auf Johann Karl Wezel und Jean Paul*. Würzburg: Königshausen & Neumann, 1989.

Kosmyna, Nataliya. "Consumer Brain-Computer Interfaces: From Science Fiction to Reality." Presentation at the Microsoft Research Lab, Redmond, WA, March 14, 2019. https://www.media.mit.edu/events/consumer-brain-computer-interfaces-from-science-fiction-to-reality/.

Kosmyna, Nataliya, Caitlin Morris, Utkarsh Sarawgi, and Pattie Maes. "AttentivU: A Biofeedback System for Real-Time Monitoring and Improvement of Engagement." In *CHI EA '19: Extended Abstracts of the 2019 CHI Conference on Human Factors in Computing Systems*, 1–2. New York: Association for Computing Machinery, 2019. http://doi.org/10.1145/3290607.3311768.

Kosmyna, Nataliya, Utkarsh Sarawgi, and Pattie Maes. "AttentivU: Evaluating the Feasibility of Biofeedback Glasses to Monitor and Improve Attention." In *Proceedings of the 2018 ACM International Joint Conference and 2018 International Symposium on Pervasive and Ubiquitous Computing and Wearable Computers*. New York: Association for Computing Machinery, 2018.

Krantz, David L. "The Baldwin-Titchener Controversy: A Case Study in the Functioning and Malfunctioning of Schools." In *Schools of Psychology: A Symposium of Papers*, ed. David L. Krantz, 1–19. New York: Appleton-Century-Crofts, 1969.

Kühnel, Jana, et al. "Staying in Touch While at Work: Relationships Between Personal Social Media Use at Work and Work-Nonwork Balance and Creativity." *International Journal of Human Resource Management* 31, no. 10 (2020): 1235–61.

Langan, Robert. "On Free-Floating Attention." *Psychoanalytic Dialogues* 7, no. 6 (1997): 819–39.

Leary, David E. "Telling Likely Stories: The Rhetoric of the New Psychology, 1880–1920." *Journal of the History of the Behavioral Sciences* 23, no. 4 (1987): 315–31.

Levy, David M. *Mindful Tech: How to Bring Balance to Our Digital Lives*. New Haven, CT: Yale University Press, 2016.

Lewis, Daniel. *The Feathery Tribe: Robert Ridgway and the Modern Study of Birds*. New Haven, CT: Yale University Press, 2012.

Liebherr, Magnus, et al. "Smartphones and Attention, Curse or Blessing? A Review on the Effects of Smartphone Usage on Attention, Inhibition, and Working Memory." *Computers in Human Behavior Reports* 1 (2020): 100005.

Lindahl, Jared R., et al. "The Varieties of Contemplative Experience: A Mixed-Methods Study of Meditation-Related Challenges in Western Buddhists." *PloS One* 12, no. 5 (2017): e0176239.

Linden, Mareta. *Untersuchungen zum Anthropologiebegriff des 18. Jahrhunderts*. Bern: Lang, 1976.

Lindsay, Grace W. "Attention in Psychology, Neuroscience, and Machine Learning." *Frontiers in Computational Neuroscience* 14 (2020): 21.

Lippard, Lucy R. "Spinning the Common Thread." In *The Precarious: The Art and Poetry of Cecilia Vicuña*. trans. Esther Allen, ed. M. Catherine de Zegher, 7–16. Hanover, NH: University Press of New England, 1997.

Lubbock, John. *Ants, Bees, and Wasps: A Record of Observations on the Habits of the Social Hymenoptera*. London: K. Paul, Trench, 1882.

Lutz, Antoine, et al. "Attention Regulation and Monitoring in Meditation." *Trends in Cognitive Sciences* 12, no. 4 (2008): 163–69.

Macpherson, Crawford Brough. *The Political Theory of Possessive Individualism: Hobbes to Locke*. Oxford: Oxford University Press, 1962.

Maes, Pattie. "From Distraction to Augmentation: Technology for Optimal Performance." Presentation at the AI for a Better World: Future of Infrastructure conference, MIT Club of Northern California, San Francisco, July 14, 2020. https://www.youtube.com/watch?v=yrHTncDuQL4.

Margaret Naumburg Papers. Kislak Center for Special Collections, Rare Books and Manuscripts, University of Pennsylvania.

Martin, Tom. "Relational Perception and 'the Feel' for Tools in the Wooden Boat Workshop." *Phenomenology & Practice* 15, no. 2 (2020): 9.

Maryland Public Television. "What Was the Underground Railroad?" *Pathways to Freedom: Maryland & the Underground Railroad*, accessed October 1, 2021. https://pathways.thinkport.org/about/about1.cfm.

Matteson, Scott. "Tapping into the Power of Thync." *Tech Republic* (blog), March 11, 2015. https://www.techrepublic.com/article/tapping-into-the-power-of-thync/.

McClelland, Tom. "The Mental Affordance Hypothesis." *Mind* 129 (2020): 401–27.

McGee, Adam Michael. "Imagined Voodoo: Terror, Sex, and Racism in American Popular Culture." PhD diss., Harvard University, 2014.

McHugh, Molly. "Heads-On with Thync, the Device That Changes Your Brain." *Daily Dot*, January 7, 2015.

McMahan, David L. *The Making of Buddhist Modernism.* New York: Oxford University Press, 2008.

Merritt, Justin. "Jhana Wars! Pt. 1 What Is Jhana Really?" *Simple Suttas* (blog), May 9, 2013. https://simplesuttas.wordpress.com/2013/05/09/jhana-wars-pt-1-what-the-heck-is-jhana-a-first-pass/.

Meyer, Steven. *Irresistible Dictation: Gertrude Stein and the Correlations of Writing and Science.* Stanford, CA: Stanford University Press, 2001.

Michaelson, Jay. *Evolving Dharma: Meditation, Buddhism, and the Next Generation of Enlightenment.* Berkeley, CA: North Atlantic, 2013.

Millar, Andrew J. "A Suite of Photoreceptors Entrains the Plant Circadian Clock." *Journal of Biological Rhythms* 18, no. 3 (2003): 217–26.

Minkowski, Eugène. *Lived Time: Phenomenological and Psychopathological Studies.* trans. Nancy Metzel. 1933; repr., Evanston, IL: Northwestern University Press, 1970.

MIT Media Lab. "Fluid Interfaces: Overview." accessed February 18, 2021. https://www.media.mit.edu/groups/fluid-interfaces/overview/.

———. "Project AttentivU: Overview." https://www.media.mit.edu/projects/attentivu/overview/, accessed February 18, 2021.

Mole, Christopher. *Attention Is Cognitive Unison: An Essay in Philosophical Psychology.* New York: Oxford University Press, 2011.

Montag, Christian, et al. "Addictive Features of Social Media/Messenger Platforms and Freemium Games against the Background of Psychological and Economic Theories." *International Journal of Environmental Research and Public Health* 16, no. 14 (2019): 2612.

Montemayor, Carlos. "Inferential Integrity and Attention." *Frontiers in Psychology* 10 (2019): 2580. https://doi.org/10.3389/fpsyg.2019.02580.

———. *Minding Time: A Philosophical and Theoretical Approach to the Psychology of Time.* Leiden: Brill, 2013.

Montemayor, Carlos and Harry H. Haladjian. *Consciousness, Attention, and Conscious Attention.* Cambridge, MA: MIT Press, 2015.

Morris, Robert. "Notes on Sculpture, Part IV: Beyond Objects." *Artforum* 7, no. 8 (1969): 54.

Moss, Stephen. *A Bird in the Bush: A Social History of Birdwatching.* London: Aurum, 2004.

Mundy, Rachel. *Animal Musicalities: Birds, Beasts, and Evolutionary Listening.* Middletown, CT: Wesleyan University Press, 2018.

Münsterberg, Hugo. *Psychological Laboratory of Harvard University.* Cambridge, MA: Harvard University Press, 1893.

Muse. "Muse with Co-founder Ariel Garten." YouTube video, 2:31, uploaded May 17, 2016. https://www.youtube.com/watch?v=GQlKaXguD2Y.

N.a. "A History of Philosophy." *Mind* 15 (1890): 132–33.

———. "Notes." *Mind* 13 (1888): 317–18.

Naumburg, Margaret. *An Introduction to Art Therapy: Studies of the "Free" Art Expression of Behavior Problem Children and Adolescents as a Means of Diagnosis and Therapy.* New York: Teachers College Press. 1973.

———. "A Study of the Art Work of a Behavior-Problem Boy as It Relates to Ego Development and Sexual Enlightenment." *Psychiatric Quarterly* 20, no. 1 (1945): 74–112.

———. *Dynamically Oriented Art Therapy: Its Principles and Practices—Illustrated with Three Case Studies.* Chicago: Magnolia Street, 1987.

North, Paul. *The Problem of Distraction.* Stanford, CA: Stanford University Press, 2011.

Olah, Chris, and Shan Carter. "Attention and Augmented Recurrent Neural Networks." *Distill*, September 8, 2016. https://distill.pub/2016/augmented-rnns/.

Olin, Margaret. *Forms of Representation in Alois Riegl's Theory of Art*. University Park, PA: Penn State University Press, 1992.

Ortner, Sherry B. "On Key Symbols." *American Anthropologist* 75, no. 5 (1973): 1338–46.

Pedersen, Morten Axel, Kristoffer Albris, and Nick Seaver. "The Political Economy of Attention." *Annual Review of Anthropology* 50, no. 1 (2021): 309–25.

Pessoa, Luiz. "On the Relationship Between Emotion and Cognition." *Nature Reviews Neuroscience* 9, no. 2 (2008): 148–58.

Petersen, Steven E., and Michael I. Posner. "The Attention System of the Human Brain: 20 Years After." *Annual Review of Neuroscience* 35 (2012): 73–89.

Pettman, Dominic. *Infinite Distraction*. New York: Polity, 2016.

Pind, Jorgen. *Edgar Rubin and Psychology in Denmark: Figure and Ground*. Dordrecht: Springer, 2014.

Porter, Theodore M. "How Science Became Technical." *Isis* 100, no. 2 (2009): 292–309.

Posner, Michael I. *Attention in a Social World*. New York: Oxford University Press, 2012.

———. "Orienting of Attention." *Quarterly Journal of Experimental Psychology* 32, no. 1 (1980): 3–25.

Posner, Michael I., and Steven E. Petersen. "The Attention System of the Human Brain." *Annual Review of Neuroscience* 13 (1990): 25–42.

Prinz, Jesse J. *The Conscious Brain: How Attention Engenders Experience*. Oxford: Oxford University Press, 2012.

Purser, Ronald E. *McMindfulness: How Mindfulness Became the New Capitalist Spirituality*. London: Repeater, 2019.

Pylyshyn, Zenon W. "Visual Indexes, Preconceptual Objects, and Situated Vision." *Cognition* 80, nos. 1–2 (2001): 127–58.

Quli, Natalie. "Multiple Buddhist Modernisms: Jhāna in Convert Theravāda." *Pacific World* 10, no. 1 (2008): 225–49.

Rabinbach, Anson. *The Human Motor: Energy, Fatigue, and the Origins of Modernity*. New York: Basic Books, 1990.

Radford, Alec, et al., "Better Language Models and Their Implications." OpenAI. February 14, 2019. https://openai.com/blog/better-language-models/.

Ramazani, Jahan. *Poetry of Mourning: The Modern Elegy from Hardy to Heaney.* Chicago: University of Chicago Press, 1994.

Reik, Theodor. *Listening with the Third Ear: The Inner Experience of a Psychoanalyst.* New York: Farrar, Straus, 1949.

Ribot, Théodule-Armand. *Psychologie de l'attention.* Paris: Germer Baillière, 1889.

Rideout, Victoria, and Michael B. Robb. *The Common Sense Census: Media Use by Tweens and Teens.* San Francisco: Common Sense Media, 2019. https://www.commonsensemedia.org/sites/default/files/research/report/2019-census-8-to-18-full-report-updated.pdf.

Riedel, Wolfgang. *Die Anthropologie des jungen Schiller: Zur Ideengeschichte der medizinischen Schriften und der 'Philosophischen Briefen'.* Würzburg: Königshausen & Neumann, 1985.

Riegl, Aloïs. "Excerpts from The Dutch Group Portrait (1902)." trans. Benjamin Binstock. *October* 74 (1995): 11.

Robbins, Christa Noel. "The Sensibility of Michael Fried." *Criticism* 60, no. 4 (2018): 432.

Roberts-Mahoney, Heather, Alexander J. Means, and Mark J. Garrison, "Netflixing Human Capital Development: Personalized Learning Technology and the Corporatization of K-12 Education," *Journal of Education Policy* 31, no. 4 (2016): 405–20.

Robinson, David Kent. "Reaction-Time Experiments in Wundt's Institute and Beyond." In *Wilhelm Wundt in History: The Making of a Scientific Psychology*, ed. R. W. Rieber and David Kent Robinson, 161–204. New York: Kluwer Academic, 2001.

Ross, Tricia M. "Anthropologia: An (Almost) Forgotten Early Modern History." *Journal of the History of Ideas* 79 (2018): 1–22.

Rubin, Edgar. *Synsoplevede figurer: studier i psykologisk analyse.* Vol. 1. Copenhagen: Gyldendalske Boghandel, Nordisk Forlag, 1915.

Sacks, Peter. *The English Elegy: Studies in the Genre from Spenser to Yeats.* Baltimore, MD: Johns Hopkins University Press, 1985.

Sadock, Benjamin James, Virginia Alcott Sadock, and Pedro Ruiz, *Kaplan & Sadock's Synopsis of Psychiatry: Behavioral Sciences/Clinical Psychiatry*, 11th ed. Philadelphia: Wolters Kluwer, 2015.

Sapolsky, Robert. "To Understand Facebook, Study Capgras Syndrome: This Mental Disorder Gives Us a Unique Insight into the Digital Age." *Nautilus*, October 27, 2016. https://nautil.us/to-understand-facebook-study-capgras-syndrome-236173/.

Saron Lab. "The Shamatha Project." accessed August 31, 2022. https://saronlab.ucdavis.edu/shamatha-project.html.

Sartre, Jean-Paul. *Being and Nothingness*. trans. H. E. Barnes. New York: Philosophical Library, 1956. First published in French as *L'être et le néant* in 1943 by Gallimard (Paris).

———. *The Imaginary: A Phenomenological Psychology of the Imagination*. trans. Jonathan Webber. London: Routledge, 2004. First published in French as *L'imaginaire* in 1940 by Gallimard (Paris).

Satell, Greg. "How Technology Enhances Creativity." *Forbes*, January 27, 2014. https://www.forbes.com/sites/gregsatell/2014/01/27/how-technology-enhances-creativity/?sh=3ff9ba6c3f50.

Schaffer, Simon. "Astronomers Mark Time: Discipline and the Personal Equation." *Science in Context* 2, no. 1 (1988): 115–45.

Scheler, Max. *The Nature of Sympathy*. trans. Peter Heath. London: Routledge, 1954.

Schiller, Friedrich. *Schillers Werke: Nationalausgabe*, ed. Benno von Wiese and Helmut Koopmann. Weimar: Hermann Böhlaus Nachfolger, 1962.

Schmidgen, Henning. "Time and Noise: The Stable Surroundings of Reaction Experiments, 1860–1890." *Studies in History and Philosophy of Science Part C: Studies in History and Philosophy of Biological and Biomedical Sciences* 34, no. 2 (2003): 237–75.

Schmidt, Benjamin MacDonald. "Paying Attention: Imagining and Measuring a Psychological Subject in American Culture, 1886-1960." PhD diss., Princeton University, 2013.

Schneewind, J. B. *The Invention of Autonomy: A History of Modern Moral Philosophy*. Cambridge: Cambridge University Press, 1998.

Schnitzler, Arthur. "Lieutenant Gustl." In *Plays and Stories: Arthur Schnitzler*. trans. Richard L. Simon, ed. Egon Schwarz. London: A&C Black, 1982.

Schooler, Jonathan W., et al. "Meta-Awareness, Perceptual Decoupling and the Wandering Mind." *Trends in Cognitive Sciences* 15, no. 7 (2011): 319–26.

Schüll, Natasha. "Afterword: Shifting the Terms of the Debate." In "Shifting Attention," ed. Rebeca Jablonsky, Tero Karppi, and Nick Seaver. Special issue, *Science, Technology, & Human Values* 47, no. 2 (2021): 360–65.

———. "Afterword: Shifting the Terms of the Debate." In "Shifting Attention," ed. Rebeca Jablonsky, Tero Karppi, and Nick Seaver. Special issue, *Science, Technology, & Human Values* 47, no. 2 (2022): 360–65.

———. "HAPIfork and the Haptic Turn in Wearable Technology." In *Being Material*, ed. Marie-Pier Boucher, Stefan Helmreich, Leila W. Kinney, Skylar Tibbits, Rebecca Uchill, and Evan Ziporyn, 70–75. Cambridge, MA: MIT Press, 2019.

Schüll, Natasha Dow. *Addiction by Design: Machine Gambling in Las Vegas*. Princeton, NJ: Princeton University Press, 2012.

———. "Data for Life: Wearable Technology and the Design of Self-Care." *BioSocieties* 11, no. 3 (2016): 317–33.

Scripture, Edward Wheeler. "Accurate Work in Psychology." *American Journal of Psychology* 6, no. 3 (1894): 427.

———. *Thinking, Feeling, Doing*. Meadville, PA: Flood and Vincent, 1895.

Seaver, Nick. "Captivating Algorithms: Recommender Systems as Traps." *Journal of Material Culture* 24, no. 4 (2019): 421–36.

Seaver, Nick. "Homo Attentus: Technological Backlash and the Attentional Subject." Presentation at the Society for Social Studies of Science Annual Meeting, New Orleans, LA, 2019.

Segundo-Ortin, Miguel, and Paco Calvo. "Are Plants Cognitive? A Reply to Adams." *Studies in History and Philosophy of Science* 73 (2019): 64–71.

Shanklin, Will. "Hands-On: Thync Mood-Changing Wearable Is Like Doing Drugs, Without All the Bad Stuff." *New Atlas*, January 9, 2015.

Shorter, Edward. *A History of Psychiatry: From the Era of the Asylum to the Age of Prozac*. Hoboken, NJ: Wiley, 1998.

Siegel, Susanna. *The Rationality of Perception*. New York: Oxford University Press, 2017.

Sigala, Marianna, and Kalotina Chalkiti. "Knowledge Management, Social Media and Employee Creativity." *International Journal of Hospitality Management* 45 (2015): 44–58.

Simon, Herbert. "Designing Organizations for an Information-Rich World." In *Computers, Communications, and the Public Interest*, ed. Martin Greenberger, 37. Baltimore, MD: Johns Hopkins University Press, 1971.

Simon, Robert I. "Great Paths Cross: Freud and James at Clark University, 1909." *American Journal of Psychiatry* 124, no. 6 (1967): 831–34.

Smith, Caleb. "Disciplines of Attention in a Secular Age." *Critical Inquiry* 45, no. 4 (2019): 884–909.

Smith, Michael E., and Alan Gevins. "Attention and Brain Activity While Watching Television: Components of Viewer Engagement." *Media Psychology* 6, no. 3 (2004): 285–305.

Sokal, Michael M. "Scientific Biography, Cognitive Deficits, and Laboratory Practice: James McKeen Cattell and Early American Experimental Psychology, 1880–1904." *Isis* 101, no. 3 (2010): 531–54.

Solomons, Leon, and Gertrude Stein. "Normal Motor Automatism." *Psychological Review* 3, no. 5 (1896): 502.

Spargo, R. Clifton. *The Ethics of Mourning: Grief and Responsibility in Elegiac Literature*. Baltimore, MD: Johns Hopkins University Press, 2004.

Sperber, Dan. *Explaining Culture: A Naturalistic Approach*. Oxford: Blackwell, 1996.

Stiegler, Bernard. "Relational Ecology and the Digital Pharmakon." *Culture Machine* 13 (2012).

———. *Taking Care of Youth and the Generations*. trans. Stephen Barker. Stanford, CA: Stanford University Press, 2010.

Stuart, Daniel M. "Insight Transformed: Coming to Terms with Mindfulness in South Asian and Global Frames." *Religions of South Asia* 11, nos. 2–3 (2017): 158–81.

———. *S. N. Goenka: Emissary of Insight*. Boulder, CO: Shambhala, 2020.

Summerfield, Christopher, and Tobias Egner. "Expectation (and Attention) in Visual Cognition." *Trends in Cognitive Sciences* 13, no. 9 (2009): 403–09.

Summit Learning. "Slavery in North America." accessed September 18, 2021. https://www.summitlearning.org/guest/focusareas/862825?fromCourseId=163087.

Susser, Daniel, Beate Roessler, and Helen Nissenbaum. "Technology, Autonomy, and Manipulation." *Internet Policy Review* 8, no. 2 (2019), https://doi.org/10.14763/2019.2.1410.

Suthor, Nicola. *Rembrandt's Roughness*. Princeton, NJ: Princeton University Press, 2018.

Sutta Central. "Numbered Discourse 9.41: 4. The Great Chapter: With the Householder Tapussa." https://suttacentral.net/an9.41/en/sujato.

Sutton, Theodora. "Digital Harm and Digital Addiction: An Anthropological View." *Anthropology Today* 36, no. 1 (2020): 17–22.

Tabor, Nick. "Mark Zuckerberg Is Trying to Transform Education. This Town Fought Back." *New York Magazine*, October 11, 2018. https://nymag

.com/intelligencer/2018/10/the-connecticut-resistance-to-zucks-summit-learning-program.html.

Taylor, Andrea Faber, and Frances E. Kuo. "Children with Attention Deficits Concentrate Better After Walk in the Park." *Journal of Attention Disorders* 12, no. 5 (2009): 402–409.

Tarnoff, Ben, and Moira Weigel. "Why Silicon Valley Can't Fix Itself." *Guardian*, May 3, 2018. https://www.theguardian.com/news/2018/may/03/why-silicon-valley-cant-fix-itself-tech-humanism.

Thompson, Evan. *Why I Am Not a Buddhist*. New Haven, CT: Yale University Press, 2020.

Thorndike, Edward L. "Animal Intelligence: An Experimental Study of the Associative Processes in Animals." *Psychological Review: Monograph Supplements* 2, no. 4 (1898): 30.

——. "Mental Fatigue." *Journal of the American Medical Association* 34, no. 12 (1900): 727.

Tiedemann, Dietrich. *Handbuch der Psychologie zum Gebrauche bei Vorlesungen*. Leipzig: Barth, 1804.

Titchener, Edward B. *Lectures on the Elementary Psychology of Feeling and Attention*. New York: Macmillan, 1908.

——. "Simple Reactions." *Mind* 4, no. 13 (1895): 74–81.

Treisman, Anne, and Garry Gelade. "A Feature-Integration Theory of Attention." *Cognitive Psychology* 12, no. 1 (1980): 97–136.

Tresch, John. "Buddhify Your Android." *Tricycle*. December 5, 2015. https://tricycle.org/trikedaily/buddhify-your-android/.

——. "Experimental Ethics and the Meditating Brain." In *Neurocultures*, ed. Francisco Ortega and Fernando Vidal, 49–68. Frankfurt: Peter Lang, 2011.

Tufekci, Zeynep. *Twitter and Tear Gas: The Power and Fragility of Networked Protest*. New Haven, CT: Yale University Press, 2017.

Turkle, Sherry. *Alone Together: Why We Expect More from Technology and Less from Each Other*. New York: Basic Books, 2011.

Uhl, Lemon L. *Attention: A Historical Summary of the Discussions Concerning the Subject*. Baltimore, MD: Johns Hopkins University Press, 1890.

Uncapher, Melina R., and Anthony D. Wagner, "Minds and Brains of Media Multitaskers: Current Findings and Future Directions." *Proceedings of the National Academy of Sciences* 115, no. 40 (2018): 9889–96.

University of California. "Ecologies of 'Mind.'" Buddhism, Mind, and Cognitive Science conference, Berkeley, March 2014.

U.S. Department of Education. "Providence School District, RI, ACS 2015–2019 Profile." Institute of Education Sciences, National Center for Education Statistics, accessed September 27, 2021. https://nces.ed.gov/programs/edge/TableViewer/acsProfile/2019.

Van Dam, Nicholas T., et al. "Mind the Hype: A Critical Evaluation and Prescriptive Agenda for Research on Mindfulness and Meditation." *Perspectives on Psychological Science* 13, no. 1 (2018): 36–61.

Van der Zande, Johan. "The Microscope of Experience: Christian Garve's Translation of Cicero's 'De Officiis' (1783)." *Journal of the History of Ideas* 59 (1998): 79–94.

Vaswani, Ashish, et al. "Attention Is All You Need." In *Advances in Neural Information Processing Systems 30*, ed. Ulrike von Luxburg et al., 5999–6009. Red Hook, NY: Curran Associates, 2017.

Vicuña, Cecilia. "Entering." In *New and Selected Poems*, trans. and ed. Rosa Alcalá, 85. Berkeley, CA: Kelsey Street Press, 2018.

Virilio, Paul. *Lost Dimension*. trans. Daniel Moshenberg. New York: Semiotext[e], 1991.

———. *The Aesthetics of Disappearance*. trans. Philip Beitchman. Los Angeles: Semiotext[e], 1991.

———. *The Information Bomb*. trans. Chris Turner. London: Verso, 2006.

Wagner, Anne Middleton. *Mother Stone: The Vitality of Modern British Sculpture*. New Haven, CT: Yale University Press, 2005.

Wallace, B. Alan. *The Attention Revolution: Unlocking the Power of the Focused Mind*. New York: Simon & Schuster, 2006.

Wang, Michelle H. "Woven Writing in Early China." *Art History* 42, no. 5 (2019): 836–61.

Watzl, Sebastian. *Structuring Mind: The Nature of Attention and How It Shapes Consciousness*. Oxford: Oxford University Press, 2017.

Weil, Simone. *Gravity and Grace*. trans. Emma Crawford and Mario von der Ruhr. London: Routledge, 2002. First published in French as *La Pesanteur et la grâce* in 1947 by Librairie PLON (Paris).

West, Martin, trans. *Homeric Hymns*. Cambridge, MA: Harvard University Press, 2003.

Wiener, Norbert. *The Human Use of Human Beings*. London: Free Association, 1989.

Williams, James. *Stand Out of Our Light: Freedom and Resistance in the Attention Economy*. Cambridge: Cambridge University Press, 2018.

Williamson, Ben. "Brain Data: Scanning, Scraping and Sculpting the Plastic Learning Brain Through Neurotechnology." *Postdigital Science and Education* 1 (2019): 65–86. https://doi.org/10.1007/s42438-018-0008-5.

Wilmer, Henry H., Lauren E. Sherman, and Jason M. Chein. "Smartphones and Cognition: A Review of Research Exploring the Links between Mobile Technology Habits and Cognitive Functioning." *Frontiers in Psychology* 8 (2017): 605.

Wisdom 2.0. "Time Well Spent: Taking Back Our Lives and Attention—Tristan Harris, Laurie Segall." YouTube video, 22:02, uploaded March 29, 2018. https://www.youtube.com/watch?v=UJ9OqzlE_zQ.

Wolff, Christian. *Vernünftige Gedanken von Gott, der Welt und der Seele des Menschen*. Halle, 1720.

Wu, Wayne. "Attention as Selection for Action." In *Attention: Philosophical and Psychological Essays*, ed. Christopher Mole, Declan Smithies, and Wayne Wu, 97–116. New York: Oxford University Press, 2011.

Xu, Kelvin, et al. "Show, Attend and Tell: Neural Image Caption Generation with Visual Attention." ArXiv:1502.03044 [Cs]. April 19, 2016. https://doi.org/10.48550/arXiv.1502.03044.

Yanni, Carla. *The Architecture of Madness: Insane Asylums in the United States*. Minneapolis: University of Minnesota Press, 2007.

Zabelina, Darya L. "Attention and Creativity." In *The Cambridge Handbook of the Neuroscience of Creativity*, ed. Rex E. Jung and Oshin Vartanian, 161–79. Cambridge: Cambridge University Press, 2018.

Zammito, John H. *Kant, Herder, and the Birth of Anthropology*. Chicago: University of Chicago Press, 2002.

Zedelius, Claire M., and Jonathan W. Schooler. "Capturing the Dynamics of Creative Daydreaming." In *Creativity and the Wandering Mind*. ed. David D. Preiss, Diego Cosmelli, and James C. Kaufman, 55–72. London: Academic, 2020.

Zhang, Maggie. "Self-Optimization Inside Classrooms: A Network Analysis on BrainCo. Educational Neurotechnology." BS, New York University, 2020.

Zuboff, Shoshana. *The Age of Surveillance Capitalism: The Fight for a Human Future at the New Frontier of Power*. New York: Public Affairs, 2019.

Zuckert, Johann F. *Von den Leidenschaften*. Berlin, 1774.

CONTRIBUTORS

Lucy Alford is an assistant professor of literature at Wake Forest University. As both a practicing poet and teacher of poetry, Alford is particularly interested in sensory life, experimentation, and the roles of habit, constraint, and play in creative processes. Her poems have appeared in *Harpur Palate*, *Literary Matters*, the *Warwick Review*, *Streetlight*, and *Atelier*. She is the author of *Forms of Poetic Attention* (Columbia University Press, 2020), which examines the forms of attention both required and produced in poetic language, bringing both philosophical and cognitive inquiry into conversation with the inner workings of specific poems.

D. Graham Burnett trained in the History and Philosophy of Science at Cambridge University and teaches at Princeton. In 2023 he was an artist in residence at the Academy of Fine Arts in Helsinki. He works at the intersection of historical inquiry and artistic practice, and focuses on experimental/experiential approaches to textual material, pedagogical modes, and hermeneutic activities traditionally associated with the research humanities. He collaborates regularly with the research collective ESTAR(SER), as well as the "Friends of Attention." Burnett edits "Conjectures" (a series of historiographical experiments) for *The Public Domain Review*, and his co-edited, co-authored volume *Twelve Theses on Attention* appeared in 2022. Recent installation projects include: "The Milcom Memorial Reading Room and Attention Library" at the Monira Foundation, Mana Contemporary (Jersey City) and *THE THIRD, MEANING* at the Frye Art Museum (Seattle). He also co-curated "Practices of Attention" for the 33rd São Paulo Biennial.

Julian Chehirian is a Ph.D. candidate in the Program for the History of Science and a fellow in the Interdisciplinary Doctoral Program in the Humanities at Princeton University. His scholarship attends to intersecting discourses on psychotherapy and attention. His dissertation project will trace the emergence of artmaking as a form of clinical psychological therapy across the twentieth century, studying the entwinement of psychoscientific and artistic practices in experiments that have used artistic expression as an inlet into the psyche. He was previously a Fulbright Researcher in Bulgaria, a Fellow at the Center for Digital Humanities at Princeton, and a Visiting Researcher at the American Research Center in Sofia.

Henry M. Cowles is a historian of science and medicine and associate professor of history at the University of Michigan. He writes and teaches about a range of topics, including psychology, addiction, self-help, and expertise.

Joanna Fiduccia is an assistant professor of the History of Art at Yale University, where she teaches and writes about European and American modernism and the historical avant-garde. She is also the author of essays and reviews on contemporary art, as well as a member of the research collective ESTAR(SER) and collaborator with the Friends of Attention. Her first book, *Figures of Crisis: Alberto Giacometti and the Myths of Nationalism*, is forthcoming.

Jonardon Ganeri is Bimal K. Matilal Distinguished Professor of Philosophy at the University of Toronto. His books include *The Self: Naturalism, Consciousness, and the First-Person Stance* (2012) and *Attention, Not Self* (2020). He is a fellow of the British Academy and a recipient of the Infosys Prize in the Humanities. He is the project leader for the Toronto/NYU-based project "Virtues of Attention."

Yael Geller Yael Geller is a psychiatrist and Ph.D. candidate at Princeton University in the Program in History of Science. Her research sits at the confluence of psychology, psychiatry, neurology, philosophy, cultural criticism, and ethics. She holds an M.D. and an M.A. in Literature from Tel Aviv University. Her first novel, *Land of Ararat*, was published in Israel in 2016.

Alexandra Hui is an associate professor of history at Mississippi State University, head of History of Science, Technology, and Medicine there, specializing in the history and psychophysics of sound, and especially of sound studies in nineteenth- and twentieth-century Germany. Among her publications are *The Psychophysical Ear: Musical Experiments, Experimental Sounds, 1840–1910* and the coedited volume, *Testing Hearing; The Making of Modern Aurality*.

Carolyn Dicey Jennings is professor of philosophy at University of California, Merced, where she researches the nature of attention and its impact on the mind (e.g., on perception, consciousness, and action). Her research has been published in *Analysis, Synthese, Journal of the American Philosophical Association,*

Neuroethics, Consciousness & Cognition, and *Philosophical Studies*. Her first book, *The Attending Mind*, was published in 2020.

Carlos Montemayor is a professor of philosophy at San Francisco State University. His research focuses on the intersection between philosophy of mind, epistemology, and cognitive science. His books include *Minding Time: A Philosophical and Theoretical Approach to the Psychology of Time*, *Consciousness, Attention and Conscious Attention* (with H. H. Haladjian), *Knowledge, Dexterity and Attention* (with Abrol Fairweather), and *The Prospect of a Humanitarian Artificial Intelligence: Agency and Value Alignment*, an open access book that illuminates the development of artificial intelligence (AI) by examining our drive to live a dignified life.

Natasha Dow Schüll is a cultural anthropologist and associate professor in the Department of Media, Culture, and Communication at New York University. Her work explores the psychic life of technology with a focus on themes of attention, anxiety, and affect modulation. She is the author of *Addiction by Design: Machine Gambling in Las Vegas* (2012), an ethnographic exploration of the relationship between technology design and the experience of addiction. Her current book project, "Keeping Track," concerns the rise of digital self-tracking technologies and the new modes of introspection and self-governance they engender.

Nick Seaver is an assistant professor of anthropology at Tufts University, where he also directs the program in Science, Technology and Society. His work studies how people use technology to make sense of cultural things. His first book, *Computing Taste: Algorithms and the Makers of Music Recommendation*, was published in 2022.

CONTRIBUTORS ◌ 347

Justin E. H. Smith is a professor of philosophy in the Department of History and Philosophy of Science at the Université Paris Cité. In 2019–2020 he was the John and Constance Birkelund Fellow at the Cullman Center for Scholars and Writers of the New York Public Library. His most recent book, *The Internet Is Not What You Think It Is*, appeared in 2022. He is currently working on a book titled *The Philosopher and the Tsar: Leibniz, Science, and the Birth of Modern Russia*, and on a fictional translation of the Voynich Manuscript. He is also translating a volume of *Olonkho*, the Siberian oral-epic tradition. He is a coeditor, with D. Graham Burnett and Catherine Hansen, of *In Search of the Third Bird: Exemplary Essays from* The Proceedings of ESTAR(SER), *2001-2021*, a work of historiographical metafiction, which appeared in 2021.

Richard J. Spiegel is a Ph.D. candidate at Princeton University in the Program for the History of Science and the Interdisciplinary Doctoral Program in the Humanities. His research focuses on European intellectual, social, and cultural history from 1700, with an emphasis on the history of the humanities and the human sciences.

Shadab Tabatabaeian obtained her Ph.D. in Cognitive and Information Sciences from University of California, Merced.

John Tresch is a professor of History of Art, Science, and Folk Practice at the Warburg Institute, School of Advanced Study, University of London. He is editor of the *History of Anthropology Review* (histanthro.org) and author of *The Romantic Machine: Utopian Science and Technology after Napoleon* (Pfizer Prize, 2012) and *The Reason for the Darkness of the Night: Edgar Allan Poe and the Forging of Modern Science* (Quinn Prize, 2021). His current book project, on images of the universe, is called *Cosmograms: How to Do Things with Worlds*.

Brian Yuan is a PhD candidate at Princeton University in the department of Anthropology. His research interests sit at the point where design, power, and energy meet in everyday technology and technological practices. His current research deals with novel formations of energy sovereignty in the United States amidst the transition to renewable energy.

INDEX

Page numbers in *italics* indicate figures or tables.

Abel, Jacob Friedrich, 31, 42n15
abhijna (psychic powers), 168
absence: as affordance, 146;
  of consciousness, 156, 299;
  epistemology of, 150; figure-
  ground structure and, 151–155,
  154; in ICUs, 294–295; of
  imagination, 157; inference
  and, 143, 149; nothingness and,
  152–153; phenomenology of,
  151–156; virtue in, 149. *See also*
  negative attention
absent-mindedness, 283, 285
Absorption and Theatricality (Fried),
  259
abstraction: attention and, 28, 29,
  33–34, 35; Herder on, 35; Kant
  on, 33–34
accommodation, 78n1
activity theorists, 121n12
actuated attention, 209n21
ADD. *See* attention deficit disorder

addiction: artificial salience and, 132,
  133, 139; in attention economy,
  180; to digital technology, 130,
  133, 189, 242; to social media, 116;
  to video gambling, 223
ADHD. *See* attention deficit
  hyperactivity disorder
Adventures of Huckleberry Finn
  (Twain), 117–118
advertising: on digital technology,
  128; industrial revolution in,
  13–14; for Muse headband, 101,
  195, *195*
aesthetic perception: imagination
  and, 145; imaginative perception
  and, 150–151, 155; negative
  attention and, 143–147
Aesthetics of Disappearance, The
  (Virilio), 289n15
Afanasyev, Alexander, 92
affordance, 146; Gestaltist
  psychology and, 147; mental,

affordance (*continued*)
  158n8; self-possession and, 222, 223, 224, 226; of Summit Learning Platform, 224
AI. *See* artificial intelligence
Aimone, Chris, 193–194
Alford, Lucy, 18, *311*, *312*
ALMS. *See* Attention Liberation Movements
alpha waves, 283
ambient becoming, 278
anapana (awareness of breath), 165, 170
anatta (soul), 168
Angell, James Rowland, 55
anicca (changing nature), 168
Anlage (experimental aptitude), 50
anthropology, 30; the Enlightenment and, 42n12; philosophy and, 33; Reformation and, 42n12
Anthropology (Kant), 33–34
Anthropology for Physicians and the Worldly (Platner), 30
anxiety: with ADD, 278, 284; with attention, 8–9, 13; dreams about, 92; FeelZing patch and, 202; mind and, 9; mindfulness for, 173; with OCD, 278, 284, 286, 288; personalized learning and, 225; with speed of life, 8–9; Summit Learning Platform and, 225
Arpaly, Nomy, 117–118
Arrival of a Train, The, 276
Arschwager, Richard, 252
art: attention and, 249–268; capitalism and, 267; free-floating attention in, 260–261; objectivity in, 262, 265; in psychotherapeutics, 94–98, *98*; self in, 259
"Art and Objecthood" (Fried), 259
artificial intelligence (AI): attention and, 230–246; context vectors in, 234–235; decoders in, 234, 235; defense from, 238—242; distraction and, 239; encoders in, 234; GPT-2, 230–233; key symbols in, 233, 238, 245–246; social media and, 231–232; thought vectors in, 235
artificial salience, 132, 133, 139
art therapy, 15
arupa (formless), 168, 178
Arvidson, P. Sven, 155, 159n20
asylums, for psychotherapeutics, 84, 88
attention: abstraction and, 28, 29, 33–34, 35; actuated, 209n21; AI and, 230–246; anxiety with, 8–9, 13; art and, 249–268; autonomy and, 105, 106, 109, 112–114, 119; biases of, 113–116; to bird sounds, 63–78; boredom and, 44–59; Buddhism and, 16, 160–181; cathexis of, 86, 87; cognitive psychology and, 40n5; competitive, 117; consciousness and, 55, 285; creativity and, 16; defined, 105–108; as dependent variable, 50, 51; digital technology and, 16, 118–120, 124–139; as discontinuous, 266–267; discovery of, 23–40;

dromology and, 276–277;
ecology of, 6–7, 50–51; in the
Enlightenment, 25, 26–29;
exogenous, 241; financialization
and, 14; flow of, 252–254; freedom
and, 35; habit and, 53, 56; hunger
and, 54–55; hypercommodified,
13; in ICUs, 18, 291–311, 313n2;
imagination and, 142–157; as
independent variable, 49–50;
inference and, 115–116; inhibition
in, 113; intelligence and, 105–108;
involuntary, 16; jhāna and,
16, 160–181; Kant on, 42n18;
in machine learning, 17; for
managing excess, 9; materiality
of, 17–18, 103–105, *104*, 108–112;
medium focus of, 254–255;
mental fatigue and, 56; mind
and, 39; mind-body reciprocity
and, 31, 34; moral philosophy
and, 15, 26; needs and, 116–117;
notation for, 71–76; ownership
of, 212–226; pain and, 28;
perceptual, 114, 115; perfection in,
28; personalized learning and,
212–226; pop-out, 150; problems
with, 13; psychoanalysis and, 13,
15, 85–91; psychotherapeutics and,
81–99; reaction time and, 49–52,
51; remote sensing experiment
and, 3–6; scientific concepts of,
125–127; selective functions of,
112–114; soul and, 29; speed of
life and, 8–9; subjectivity of, 53;
unfamiliarity and, 118–120; for
vigil, 18; virtue and, 103–104, 114,

115–116, 117; volition as, 84–85,
86; wearable technology and,
187–207, *188*. *See also* bottom-up
attention; free-floating attention;
involuntary attention; negative
attention; top-down attention;
voluntary attention
attentional crisis, 8, 10–11
attentional regime, 191
attentional sovereignty, 241; with
wearable technology, 191, 204,
206–207, 208n9
attention deficit disorder (ADD), 18,
275–288; anxiety with, 278, 284;
in attention economy, 287–288;
speed of life with, 275–277
attention deficit hyperactivity
disorder (ADHD), 284–285
attention economy, 232–233, 247n4,
287–288; addiction in, 180
Attention Liberation Movements
(ALMS), 14
Attention Revolution, The
(Wallace), 161
attention span: capitalism and,
266–267; with digital technology,
242; involuntary attention and,
283
attentive listening: accommodation
and, 78n1. *See also* bird sounds
attentive subject, 254; Fried on, 260
AttentivU, 187, *188*, 190, 196–199, *197*,
205, 210n24
auditory memory, of bird sounds,
71, 77, 78
Auk, The (Stone), 66
automatic writing experiments, 53–54

automatism, 53–54, 267
autonomic nervous system, 200
autonomy: attention and, 105, 106, 109, 112–114, 119; in culture, 105; digital technology and, 126, 133; for intelligence, 108; top-down attention and, 126
Avis Tertia. See Order of the Third Bird
awareness of breath (*anapana*), 165, 170

Bahdanau, Dzmitry, 236–237
Baldwin, James Mark, 50
barred owl, 68, 70
Baumgarten, Alexander, 25, 28, 33–34
Baxandall, Michael, 257
behaviorism, 40n5, 110–112, 121n9
Being and Nothingness (Sartre), 151–153
Bekhterev, V. M., 110
Beneke, Friedrich, 37, 38
benevolence (*metta*), 163, 170, 180–181
Bengio, Yoshua, 237
Bermúdez, Pablo, 132
Besonnenheit (reflective consciousness and deliberateness), 34
Bhattacharyya, Krishnachandra, 16, 142–157
Bhikku, Thanissaro, 179
biases: of attention, 113–116; of digital technology, 132, 133–134
Bill & Melinda Gates Foundation, 214–215, 220
binding problem, 113–114
Bird-Lore (Stone), 66

bird sounds: attention to, 63–78; auditory memory of, 71, 77, 78; barred owl, 68, 70; goshawk, 73; Nashville warbler, 72, 77; notation for, 64, 67–69, *68, 69*, 71–76; olive-sided flycatcher, 75–76; white-crowned sparrow, 68, *68–69, 69, 73*, 73–75; yellow-throated vireo, 63–78, *64*
Birds. See Order of the Third Bird
Bleuler, Eugene, 278
Boilly, Louis-Léopold, 103, *104*
boredom: attention and, 44–59; AttentivU and, 198; ennui and, 46; flow of, 45; isolation and, 57–58, 57–59; with jhāna, 177; in mind, 47; mind-wandering and, 45
Boring, Edwin, 15, 56
bottom-up attention, 16, 115, 116, 125–126; in creativity, 134–135; creativity and, 137, 138; digital technology and, 132; mind-wandering and, 135–136
Box with the Sound of Its Own Making (Morris), 17–18, 249–252, *250–251, 253*
Brahm, Ajahn, 178
brain: attention and, 125; AttentivU and, 196–197, 198; creativity and, 126, 135–136; digital technology and, 131; FeelZing patch and, 200–201; Freud and, 99n8; jhāna and, 161, 176; wearable technology and, 190
brain-machine interface, 17
Brasington, Leigh, 161, 163, 164, 165, 168, 170, 171–172, 174, 176, 178–181

Brenninkmeijer, Joanna, 209n14
Britton, Willoughby, 165
Brock, André, 225–226
Brothers Grimm, 93
Brown, Judith, 268
Buddhify, 175
Buddhism: modernity in, 173–176; in postmodernism, 174. *See also* jhāna
Buddhist Geeks, 175
Burtch, Verdi, 66–78, *73*, *74*

Cage, John, 249–252
"Call to Minimize Distraction and Respect Users' Attention, A" (Harris), 239
camera obscura, 11, 12
capitalism: art and, 267; attention span and, 266–267; digital technology in, 173; jhāna and, 180; surveillance, 13–14, 119
Catherine, Shaila, 178
cathexis, 86, 87
Cattell, James McKeen, 45, 53–54
causality: for absence, 153; for addiction, 132; for self-control, 129; in technological determinism, 10
Center for Humane Technology, 239–241, *240*, 244
central nervous system, 283
changing nature (*anicca*), 168
Charlesworth, Jonathan, 202–203
Chehirian, Julian, 13, 15
Chicago School, 50–52
Cho, KyungHyun, 237–238
Chomp, 121n9

Chomsky, Noam, 107, 112
Chopin, Frédéric, 63–78, *65*
Citton, Yves, 6–7, 191
clarity: of knowledge, 29; from Thync, 200; in *Vorstellungen*, 27–28
classical models, of subjectivity, 11, 12–13
click economy, 189
cognitive power (*Erkenntniskraft*), 31
cognitive psychology, attention and, 40n5
coherency, of Gestaltists, 156, *156*
coma, 7, 18, 291–309
comparatio (comparison), 28
competitive attention, 117
computational psychology, 112–113, 121n9
concentration (*samadhi*), 161
consciousness: absence of, 156, 299; attention and, 55, 285; boredom and, 46; field of, 154; figure-ground structure and, 154–155; limitless, 168; memory and, 87; negative attention and, 150–151; nonperceptual, 150; poetry and, 295; in psychoanalysis, 87; unity of, 30
conscious nonperception, 149–150
contemplative development, jhāna and, 165
context vectors, in AI, 234–235
Cooke, Edward, 265
Cornish, Mary, 298–300
Cowles, Henry M., 83
Crary, Jonathan, 10–11, 19n1, 40n6, 213, 224

creative cognition, 126, 135–137
creativity: attention and, 16, 124–139; bottom-up attention and, 134–135; digital technology and, 133–134, 136, 137–139; evaluation process of, 126, 134; exogenous cues for, 125; generation process of, 126, 134; mind-wandering and, 135–136; scientific concepts of, 125–127; technology and, 124–139; top-down attention and, 134–135
culture: ADHD and, 285; autonomy in, 105; of jhāna, 16; key symbols in, 233, 238, 245–246; unconscious and, 116. *See also* anthropology; *specific types and topics*

Darwin, Charles, 280
Daston, Lorraine, 48
Dean, James, 1
Dear Friend (Kanevsky), *282*
decoders, in AI, 234, 235
default mode network (DMN), 135–136
Deleuze, Gilles, 270n9
denial-of-attention attacks, 232
desire, 27–28
Desjarlais, Robert, 222–223
De Staalmeesters (Rembrandt), 263–265, *264*, 265
detox programs, for digital technology, 189
Dewey, John, 51–52, 94
dhamma, the *sangha* (the enlightened one, his teachings, and the community of practitioners), 163
Dharma Overground, 174
digital attention crisis, 242
digital health industry, 190, 208n6
digital technology: adaptability of, 129; addiction to, 130, 133, 189, 242; advertising on, 128; attention and, 16, 118–120, 124–139; attention span with, 242; autonomy and, 126, 133; in capitalism, 173; creativity and, 133–134, 136, 137–139; design elements of, 128–129; detox programs for, 189; distraction and, 189, 190; interactions on, 128–129; intermittent variable rewards in, 128, 132; jhāna and, 180; long-term effects of, 130–131; measurable impact of, 129–131; mind-wandering and, 136; modeling and evaluating impact of, 131–134; multitasking with, 130; needs in, 116–117; personality and, 129; self-control and, 129–130; unfamiliarity from, 118–120. *See also* artificial intelligence; social media; wearable technology
Dilthey, Wilhelm, 37
directed thought (*vitakka*), 167
distinctness, 27
distraction, 84–85; AI and, 239; of digital technology, 131; digital technology and, 189, 190; James on, 85; Muse headband and, 192–194; in personalized learning,

INDEX ☙ 355

224–225; in psychoanalysis, 85, 91, 92–94; virtue and, 104
DMN. *See* default mode network
Dobrynin, N. F., 103, 110–112; as activity theorist, 121n12
Dolven, Jeff, 7
Doty, Mark, 310
Dove, Rita, 307–308
Drama of Ideas, The (Puchner), 7
dreams, psychoanalysis of, 81–83, *82*, 92–94
Dreyfus, Hubert, 146–147
drive theory, 86
dromology, attention and, 276–277
duck-rabbit, 150
dukkha (unsatisfactoriness), 168

Ebbinghaus, Hermann, 39
ecology, of attention, 6–7, 50–51
EEG. *See* electroencephalography
efficiency, 49
ego dystonic, 287
ego syntonic, 287
eight-fold noble path, 166
elective affinities, 173
electroencephalography (EEG): in ICUs, 291; on jhāna, 176; Muse headband and, 192, 201; on television, 283
Emergent Phenomenology Research Consortium, 175
empiricism, 29, 37; of Buddhism, 174; Kant and, 36
encoders, in AI, 234
endogenous cues, 125
the enlightened one, his teachings, and the community of practitioners (*dhamma*, the *sangha*), 163
enlightenment, from jhāna, 166, 168
Enlightenment, the, 11, 14; anthropology and, 42n12; attention in, 25, 26–29; meaning of, 27; subjectivity of, 260
ennui, boredom and, 46
epistemology, of absence, 150
Erdheim, Mario, 89
Erdmann, Johann, 37
Erkenntniskraft (cognitive power), 31
ESTAR(SER), or "Esthetical Society for Transcendental and Applied Realization (now incorporating the Society of Esthetic Realizers)," 19.
evaluation (*vicara*), 167
evaluation process, of creativity, 126, 134
executive function, creativity and, 127
existential phenomenology, 16
exogenous attention, 241
exogenous cues, 125
experimental aptitude (*Anlage*), 50
experimental psychology, Ribot and, 108–109
exploration bias, 132, 133–134

Facebook, 224, 242
familiarization, 119
Fanon, Frantz, 225
Fantaissie-Impromptu (Chopin), 64–65, *65*, 77
Fechner, Gustav, 46–47, 56
Feder, Johann Georg, 29

FeelZing patch, 190, 199–203, 200, 203, 205
Fiduccia, Joanna, 17–18
Field Book of Wild Birds and Their Music (Mathews), 63–66, 77
field notebook listening. *See* bird sounds
field of consciousness, 154
figure-ground structure, absence and, 151–155, *154*
financialization, attention and, 14
fire (*jhāyati*), 166
five hindrances, 170
Flaubert, Gustave, 46
flow: of attention, 252–254; of boredom, 45; narrative, 222–224
Fluid Interfaces, at MIT, 198
focus areas, in personalized learning, 218–220
Foley, Helen, 314nn11–12
form (*rupa*), 168
formless (*arupa*), 168, 178
4'33 (Cage), 250–251
Fowler, Geoffrey A., 200, 200–201, 202
FPN. *See* frontoparietal control network
free association, 15, 88–90, 96
freedom, 28; attention and, 35; of mind, 29
free-floating attention: in art, 260–261; Freud and, 15, 90; mind-wandering and, 91; Naumburg on, 94–96; Reik on, 91
Freud, Sigmund: brain and, 99n8; dreams and, 81–83, *82*, 92–94; free-floating attention and, 15,

90; *as gleichschwebend*, 88–91; *The Interpretation of Dreams* by, 82, 87, 88; on memory, 87; *Project for a Scientific Psychology* by, 85–86. *See also* ego
Fried, Michael, 259–260, 262
Friends of Attention, 13–14, 19n2
Fries, Jakob, 37, 38
"From Distraction to Augmentation" (Maes), 198, 210n24
frontoparietal control network (FPN), 135–136

Galison, Peter, 48
gamma waves, 283
Ganeri, Jonardon, 13, 16, 183n23
Ganesh, Maya Indira, 241, 242
Garten, Ariel, 192–193
Garve, Christian, 29
Gaut, Berys, 126
Geller, Yael, 8, 18
generation process, of creativity, 126, 134
Geoff, Ajahn, 179
Gestaltists, coherency of, 156, *156*
Gestalt psychology, 40n5, 146–147
Gibson, J. J., 146
Ginzburg, Carlo, 91; on dreams, 92–93
gleichschwebend, Freud as, 88–91
Gleig, Ann, 174
Goenka, S. N., 165
Golden Rule, 1–2
Goodstein, Elizabeth, 46
goshawk, 73
GPT-2, 230–233

Gradient Learning, 214, 227n7
Greenaway, Peter, 97
grief, in vigil, 299–300, 303, 305–308
Grundriss der Geschichte der Philosophie (Erdmann), 37
Guattari, Félix, 270n9
Guenther, Katja, 99n8, 121n9
Gurwitsch, Aron, 156

habit: attention and, 53, 56; in behaviorism, 110, 111; as virtue, 104
Haller, Albrecht von, 30
happiness (*sukka*), 167–168
Harris, Tristan, 239–240, 242
Headspace, 175
Hegel, Georg Wilhelm Friedrich, 38, 260
Hepworth, Barbara, 255–258, 256, 269n6
Herbart, Johann, 14, 37, 38
Herder, Johann Gottfried von, 33, 34–35
highest self, 243
historical experience, 35
History of Experimental Psychology, A (Boring), 56
hoarding, in OCD, 286
Hollow Form with White (Hepworth), 255–258, 256, 269n6
Holly, Michael Ann, 255
Homer, 300–309, 314nn11–12
Huckleberry Finn (fictional character), 117–118
Hui, Alexandra, 15
hunger: attention and, 54–55; for materiality, 286

Hunter, Ian, 36
Hylan, John Perham, 271n25
Hymn to Demeter (Homer), 304–309, 314nn11–12
hypercommodified attention, 13
hyperkinetic impulse disorder, 284–285
hypnosis, 83; jhāna as, 179

ICUs. *See* intensive care units
idealism: of rationalism, 37; transcendental, 38
images: in absence, 150; in unconscious, 95–96. *See also* art
imagination: absence of, 157; aesthetic perception and, 145; attention and, 142–157; in *Hollow Form with White*, 255; memory and, 146
imaginative perception, 144, 145–146; aesthetic perception and, 150–151, 155; transformation in attending of, 147–151
"Imbecile Wolf, The" (Afanasyev), 92
imperfection, 27
industrial revolution, 13–14
inference: absence and, 143, 149; attention and, 115–116
information overload, 9
Ingold, Tim, 258
Ingram, Daniel, 174–176
inhibition: in attention, 113; in creativity, 134; inference and, 115; James on, 85
instrumental splitting, of vision, 283

intelligence: attention and, 105–108; autonomy for, 108; defined, 105–108
intensive care units (ICUs): absence in, 294–295; attention in, 18, 291–311, 313n2; poetry and, 295–311
intermittent variable rewards, in digital technology, 128, 132
Interpretation of Dreams, The (Freud), 82, 87, 88
involuntary attention, 16, 125; attention span and, 283; in reflexology, 109; wearable technology and, 191, 208n8
Irwing, Karl Franz von, 29
isolation, boredom and, 57–58, 57–59

James, William, 45, 46–48, 50, 56, 208n8, 271n25; on distraction, 84–85; Dobrynin and, 111; field of consciousness of, 154; on memory, 87; *The Principles of Psychology* by, 285; psychoanalysis and, 87; "The Will to Believe" by, 48
Jennings, Carolyn Dicey, 15–16
jhāna: attention and, 16, 160–181; boredom with, 177; brain and, 161, 176; capitalism and, 180; contemplative development and, 165; controversy with, 176–179; culture of, 16; digital technology and, 180; dropping in, 162–166; EEG on, 176; enlightenment from, 166, 168; form and formless of, 166–169; glimpses of, 170–173; as hypnosis, 179; mind and, 161; 166–167; mind-wandering in, 163; pleasure and, 177, 180
jhāyati (fire), 166
Jim (fictional character), 117–118
joy (*sukka*), 167–168
Judd, Donald, 260

Kanevsky, Alex, 281–282, *282*
Kant, Immanuel, 14, 32–34, 35; on attention, 42n18; philosophy after, 36–40
key symbols, 233, 238, 245–246
Khan Academy, 219
Khema, Ayya, 171
Kos'myna, Nataliya, *188*, 196–199

La Mettrie, Julien Offray de, 106–108
"Lane, The" (Cornish), 298–300
Lange, N. N., 110
Leibniz, Gottfried Wilhelm, 27, 107
libidinal energy, 86
Library of Philosophy, 37
Lieutenant Gustl (Schnitzler), 88–89
Life Histories of North American Birds (Stone), 66
limitless consciousness, in jhāna, 168
limitless space, in jhāna, 168
Lindahl, Jared, 165
Listening with the Third Ear (Reik), 91
Localization and Its Discontents (Guenther), 121n9
lLoge, Une (Boilly), 103, *104*
Lubbock, John, 54
Lumière brothers, 276
Luther, Martin, 175

machine learning: attention in, 17.
  See also artificial intelligence
Maes, Pattie, 198, 210n24
Majjhima Nikaya (Buddha), 176
Man a Machine (La Mettrie), 106, 107
Man a Plant (La Mettrie), 106, 107
Manichaean devils, 248n16
Marr, David C., 112
Martin, Tom, 267
Mastering the Core Teachings of Buddhism (Ingram), 174–175
materiality: of attention, 17–18, 103–105, *104*, 108–112; hunger for, 286; medium focus and, 255; of mind, 109; OCD and, 286–287
Mathews, F. Schuyler, 63–66, *64*, *68*, 76–77
McIlvaine, J. H., 83–84
McMahan, David, 173
mechanical objectivity, 48, 60n6
meditation: Muse headband and, 192–195; pain from, 177; Vipassana, 161, 165, 169, 171, 177. See also jhāna; mindfulness
Meditations on Emptiness and Dependent Arising, 171
medium focus, of attention, 254–255
Meier, Georg, 29
memory: auditory, of bird sounds, 71; consciousness and, 87; creativity and, 127, 135; Freud on, 87; imagination and, 146; James on, 87; solicitations from, 89
Mendelssohn, Moses, 29
mental affordance, 158n8

mental fatigue, 56
mental representations (*Vorstellungen*), 27
Metaphysics (Baumgarten), 28, 33–34
metta (benevolence), 163, 170, 180–181
Michaelson, Jay, 174
mind: anxiety and, 9; attention and, 39; boredom in, 47; camera obscura and, 11; in coma, 293; freedom of, 29; jhāna and, 161, 166–167; materiality of, 109; mechanical view of, 107; soul and, 34
Mind, 37
mind-body reciprocity, 31, 34
mindfulness, 161; for anxiety, 173; technology and, 16
Mind & Life, 164
mind-wandering: AttentivU and, 198; boredom and, 45; bottom-up attention and, 135–136; creativity and, 135–136; digital technology and, 136; free-floating attention and, 91; in jhāna, 163; Muse headband and, 192, 194, 210n24; top-down attention and, 135–136
Minkowski, Eugène, 277–278
missingness, 149
MIT, Fluid Interfaces at, 198
modernity: in Buddhism, 173–176; psychic damage from, 84; speed of life of, 8; technology of, 8
Montemayor, Carlos, 15
Moore, Addison W., 55

moral philosophy, 15, 26
moral therapy, 83
Morris, Robert, 17–18, 249–252, *250–251*, *253*, 258–259, 260
Mother Love (Dove), 307–308
motoric theories, 110
mourning, poetry of, 295–311
multitasking, with digital technologies, 130
Münsterberg, Hugo, 47, *48*
Muse headband, 190, 192–196, *193*, *195*, 205; advanced iterations of, 208n12, 209n13; mind-wandering and, 192, 194, 210n24

narrative flow, 222–224
Nashville warbler, 72, 77
naturalism, 107
Naumburg, Margaret, 94–96
Necker cube, 150
needs, 107–108, 114; attention and, 116–117
negative attention: aesthetic perception and, 143–147; attention and, 16, 142–157; consciousness and, 150–151
negative attention (absence), transformation in attending of, 147–151
neither perception nor nonperception, in jhāna, 168
neostoic philosophy, 28, 40n10
nerve fluid, 31
nervous system: autonomic, 200; central, 283; Muse headband and, 200, 201; soul and, 30–31. *See also* brain

"Neural Machine Translation by Jointly Learning to Align and Translate," 234–235
neural networks. *See* artificial intelligence
new psychology, 45, 56
Newton, Isaac, 107
Night Watch, The (Rembrandt), 263–265, *264*
nimitta (sign), 171, 178
nonperceptual consciousness, 150
notation: for attention, 71–76; for bird sounds, 64, 67–69, *68*, *69*, 71–76
"Note on the Division of Labor by Sex, A" (Brown), 268
nothingness: absence and, 152–153; in jhāna, 168; in vigil, 292

objectivity: in art, 262, 265; isolation and, 57; mechanical, 48, 60n6; reaction times and, 49
O'Brien, Rachael, 163, 164, 171–172, 179
obsessive-compulsive disorder (OCD), 18, 275–288; anxiety with, 278, 284, 286, 288; in attention economy, 287–288; hoarding in, 286; materiality and, 286–287; materiality in, 286–287; primary slowness with, 286; speed of life with, 275–277, 286–287
Odyssey (Homer), 301–307
olive-sided flycatcher, 75–76
"On the Cognition and Sensation of the Human Soul" (Herder), 35

OpenAI, 230–233
optical illusions, 12
Order of the Third Bird, 19
orienting response, 283
Ortner, Sherry, 233, 238, 243, 246

pacing line, in individualized learning, 220, 221
pain: attention and, 28; from imperfection, 27; of loss, 295; from meditation, 177; Summit Learning Platform and, 225; video gambling and, 223
Pal, Sumon, 201
Pankejeff, Sergei, 81–82, *82*, 92–94
passion, 27–28
Path of Purification (*Visuddhimagga*), 178–179
perceptual attention, 114, 115
perfection: in attention, 28; Wolff on, 27
"Persephone in Hell" (Dove), 309
personality, 12; digital technology and, 129; in ICUs, 294
personalized learning, 212–226; anxiety and, 225; distraction in, 224–225; focus areas of, 218–220; perils of, 222–226
phenomenology: of absence, 151–156; existential, 16; in psychology, 278
philosophical physiology, 30
philosophy, 29–36; anthropology and, 33; after Kant, 36–40; moral, 15, 26. *See also specific individuals*
Philosophy of Physiology (Schiller), 32
piti (rapture), 167, 170, 177–178
Platner, Ernst, 14, 30–32, 34

pleasure: desire and, 28; jhāna and, 177, 180; from perfection, 27
poetry: consciousness and, 295; *Hymn to Demeter*, 304–309; ICUs and, 295–311; *Still Life with Oysters and Lemon*, 310; "The Lane," 298–300; "Tomb Painting," 300
pop-out attention, 150
Porter, Theodore, 48
postmodernism, Buddhism in, 174
preconscious, in psychoanalysis, 86
primary slowness, with OCD, 286
Principles of Psychology, The (James), 285
Project for a Scientific Psychology (Freud), 85–86
psychic powers (*abhijna*), 168
psychoanalysis: attention and, 13, 15, 85–91; consciousness in, 87; distraction in, 85, 91; of dreams, 81–83, *82*, 92–94; free association in, 15, 88–90, 96; James and, 87; patient-analyst relationship in, 90–91; preconscious in, 86; unconscious in, 86, 89–91, 94–96. *See also* Freud, Sigmund
psychology: boredom in, 44–59; normal in, 284–287; phenomenology in, 278; scientific, 39. *See also* attention deficit disorder; obsessive-compulsive disorder
psychotherapeutics: art in, 94–99, *98*; asylums for, 84, 88; attention and, 81–99
Puchner, Martin, 7

rapture (*piti*), 167, 170, 177–178
rationalism: idealism of, 37; Kant and, 36
Ravikovitch, Dahlia, 281
reaction time, 49–52, *51*, 60n7
reactors, 50
realism, 37
Rebel Without a Cause, 1
Red Studio (Cornish), 298–300
reflection (*reflexio*), 28
reflection, soul and, 34
reflective consciousness and deliberateness (*Besonnenheit*), 34
"Reflex Arc Concept in Psychology" (Dewey), 51–52
reflexio (reflection), 28
reflexology, 108–109
Reformation, anthropology and, 42n12
Reid, Thomas, 25
Reik, Theodor, 91
Rembrandt, 262–266, *264*, 271n23; subjectivism and, 262
remote sensing experiment, 3–6
Ribot, Théodule-Armand, 108–109, 114
Riegl, Aloïs, 260–266
right concentration (*sama samadhi*), 166
Romanticism: Buddhism and, 174; unconscious of, 89
Rubin, Edgar, 153–154, *154*
rupa (form), 168

Sacks, Peter, 314n8
samadhi (concentration), 161
sama samadhi (right concentration), 166

Sanders, Mr., 1–7
Saron, Cliff, 161, 164–165
Sartre, Jean-Paul, 16, 151–153, 155, 159n24
Sayadaw, Pa-Auk, 178
Scenes of Instruction in Renaissance Romance (Dolven), 7
Scheler, Max, 158n7
Schiller, Friedrich, 31–33
schizophrenia, 278
Schmidt, Eric, 287–288
Schniztler, Arthur, 88–89
Scholasticism, 29
Schüll, Natasha Dow, 10, 17, 222–223, 225
scientific psychology, 39
Scientific Reports, 280
scientism, Kant and, 36
Scripture, Edward, 57–59
Seaver, Nick, 10, 17, 208n9
self: in art, 259; highest, 243; locus and character of, 12; shame and, 225
self-control, digital technology and, 129–130
self-discipline, 84
self-help, James and, 48
self-improvement, 49
self-proprietorship, 17, 216
sensorial unity, 283
sequence-to-sequence learning, 234
"Seven Little Kids, The" (Brothers Grimm), 93
Shamatha Project, 161, 164
shame, 225
"Show, Attend, and Tell," 236
sign (*nimitta*), 171, 178

Simon, Herbert, 247n4
Simone, Nina, 245
Skinner, B. F., 121n9
smartphones. *See* digital technology
Smith, Caleb, 84
Social Dilemma, The, 244–245
social media, 116–117; addiction to, 116; AI and, 231–232; Facebook, 224, 242; Twitter, 124, 136, 137
sola scriptura, 175
solicitations: distraction and, 85; Gestaltists and, 147; from memory, 89; of missingness, 149
"Some Aspects of Negation" (Bhattacharyya), 142–143
soul: attention and, 29; communion with body, 31; mind and, 34; nervous system and, 30–31; reflection and, 34; Schiller on, 32; vigil and, 297; Wolff on, 27
soul (*anatta*), 168
Spätrömische Kunstindustrie (Riegl), 260
speed of life: with ADD, 275–277; anxiety with, 8–9; of modernity, 8; with OCD, 275–277, 286–287; Virilio on, 18
Spiegel, Richard J., 10, 14
standardization, 48
"Stand Out of Our Light" (Williams), 23, 239
Stein, Gertrude, 53–54
Stiegler, Bernard, 217, 219, 279
Still Life with Oysters and Lemon, 310
Stone, Clarence, 63, 66–78, *68*, *69*, *73*, *74*

Strother School of Radical Attention, 19
Subject as Freedom, The (Bhattacharyya), 143–144
subjectivism: Kant and, 36; Rembrandt and, 262
subjectivity: of attention, 53; classical models of, 11, 12–13; of the Enlightenment, 260; locus of, 13; reaction times and, 49; transcendental conditions of, 36
sukka (joy, happiness), 167–168
Sulzer, Johann, 29, 34–35
Summit Learning Platform, 17; affordance of, 224; critics of, 227n7; distraction in, 224–225; focus areas of, 218–220; pacing line in, 220, 221; perils of, 222–226; personalized learning of, 212–226
surveillance capitalism, 13–14, 119
Suspensions of Perception (Crary), 10–11, 19n1, 40n6, 213, 224
Suthor, Nicola, 271n23
systematic philosophy, 26–27

Tabatabaeian, Shadab, 15–16
Tarnoff, Ben, 239
technological determinism, 9–10; with personalized learning, 222
technology: of brain-machine interface, 17; creativity and, 124–139; mindfulness and, 16; of modernity, 8. *See also* digital technology; wearable technology

television: advertising on, 128; multitasking with, 130; orienting response with, 283
theory of distraction, 84
theory of mind, in psychoanalysis, 85
Thomasius, Christian, 26
Thompson, Evan, 161–162
Thorndike, Edward, 45, 46, 54–56, 55
thought vectors, in AI, 235
Thync, 210n27; clarity from, 200. *See also* FeelZing patch
Tiedermann, Dietrich, 29
Time Well Spent, 13–14, 239
Titchener, Edward, 23–26, 36, 37, 39–40, 40n5, 50, 53, 56
"Tomb Painting" (Cornish), 300
top-down attention, 16, 115, 116, 125–126; in creativity, 134–135; creativity and, 136, 137, 138; digital technology and, 127, 132, 133; mind-wandering and, 135–136
transcendental idealism, 38
transformation in attending, of negative attention, 147–151
Tresch, John, 16
Tufekci, Zeynep, 232
Turing, Alan, 112, 121n9
Twain, Mark, 117–118
Twitter, 124, 136, 137

unconscious: attentive biases in, 116; culture and, 116; images in, 95–96; Naumburg and, 94–96; in psychoanalysis, 86, 89–91, 94–96; of Romanticism, 89

unfamiliarity: with absence of consciousness, 299; from digital technology, 118–120
unsatisfactoriness (*dukkha*), 168

Varela, Francisco, 164
vicara (evaluation), 167
Vicuña, Cecilia, 297, 303
video gambling, 223
vigil: attention for, 18; grief in, 299–300, 303, 305–308; nothingness in, 292; soul and, 297. *See also* intensive care units
Vimalaramsi, Bhante, 171–172, 179
Vipassana meditation, 161, 165, 169, 171, 177
Virilio, Paul, 18, 275, 276–277, 281; *The Aesthetics of Disappearance* by, 289n15; *The Vision Machine* by, 282–283
virtue: in absence, 149; attention and, 103–104, 114, 115–116, 117; defined, 120n3
vision: AI and, 236; AttentivU and, 189; of emergent phenomena, 175; instrumental splitting of, 283; as stable, 12; with Summit Learning Platform, 220
Vision Machine, The (Virilio), 282–283
Visuddhimagga (*Path of Purification*), 178–179
vitakka (directed thought), 167
volition: as attention, 84–85, 86; Freud on, 86; James on, 84–85

voluntary attention, 16, 125; in reflexology, 109; with wearable technology, 191, 208n8
Vorstellungen (mental representations), 27

Wagner, Anne Middleton, 257–258
Walk Through H, A (Greenaway), 97
Wallace, Alan, 161, 164
wearable technology, 187–207; attentional sovereignty with, 191, 204, 206–207, 208n9; AttentivU, 187, *188*, 190, 196–199, *197*, 205, 210n24; brain and, 190; FeelZing patch, 190, 199–203, *200*, *203*, 205; involuntary attention and, 191, 208n8; Muse headband, 192–199, *193*, *195*, 205, 208n12, 209n13; voluntary attention with, 191, 208n8
Weber, Max, 46–47, 173

Weigel, Moira, 239
Weil, Simone, 297
Westermann, H. C., 252
white-crowned sparrow, *68*, 68–69, *69*, 73, *73*, 73–75, *74*
Wiener, Norbert, 248n16
Williams, James, 13, 23, 239, 240–241
"Will to Believe, The" (James), 48
Wilson, E. O., 240
Wolff, Christian, 26–29, 33, 35, 36
Wu, Tim, 13
Wundt, Wilhelm, 24, 37, 38, 46–47, 50, 56, 261; reaction time and, 49, 60n7

yellow-throated vireo, 63–78, *64*, 77
YouTube, 218, 219, 223, 224, 225
Yuan, Brian, 17

Zuboff, Shoshana, 13, 119
Zuckerberg, Mark, 17, 214

GPSR Authorized Representative: Easy Access System Europe, Mustamäe tee 50, 10621 Tallinn, Estonia, gpsr.requests@easproject.com

www.ingramcontent.com/pod-product-compliance
Lightning Source LLC
Chambersburg PA
CBHW022026290426
44109CB00014B/770